DINOSAUR EXTINCTION AND THE END OF AN ERA

Critical Moments in Paleobiology and Earth History Series
David J. Bottjer and Richard K. Bambach, Editors

George R. McGhee, Jr., *The Late Devonian Mass Extinction: The Frasnian/Famennian Crisis*

Betsey Dexter Dyer and Robert Alan Obar, *Tracing the History of Eukaryotic Cells: The Enigmatic Smile*

Donald R. Prothero, *The Eocene-Oligocene Transition: Paradise Lost*

Douglas H. Erwin, *The Great Paleozoic Crisis: Life and Death in the Permian*

Mark A. S. McMenamin and Dianna Schulte McMenamin, *The Emergence of Animals: The Cambrian Breakthrough*

The Perspectives in Paleobiology and Earth History Series
David J. Bottjer and Richard K. Bambach, Editors

Anthony Hallam, *Phanerozic Sea-Level Changes*

The concept for these series was suggested by Mark and Dianna McMenamin, whose book *The Emergence of Animals* was the first to be published.

Dinosaur Extinction and the End of an Era

WHAT THE FOSSILS SAY

J. DAVID ARCHIBALD

COLUMBIA UNIVERSITY PRESS
NEW YORK

Columbia University Press
New York Chichester, West Sussex

Library of Congress

Archibald, J. David.
Dinosaur extinction and the end of an era : what the fossils say / J. David Archibald.
p. cm. — (Critical moments in paleobiology and earth history series)
Includes bibliographical references and index.
ISBN 0-231-07624-X (alk. paper), — ISBN 0-231-07625-8 (pbk.)
1. Cretaceous–Tertiary boundary. 2. Extinction (Biology)
I. Title. II. Series.
QE734.5.A73 1996
567.9'1—dc20 95-31113

Casebound editions of Columbia University Press books are printed on permanent and durable acid-free paper.

Printed in the United States of America

c 10 9 8 7 6 5 4 3 2 1
p 10 9 8 7 6 5 4 3 2 1

To my colleagues, in the United States and abroad, who toil on Cretaceous faunas and floras. It is their work that reveals the patterns of evolution and extinction at the end of an Era.

For my friend and colleague Lev Alexandrovich Nessov (1947–1995)

A Rock, A River, A Tree
Hosts to species long since departed,
Marked the mastodon,
The dinosaur, who left dried tokens
Of their sojourn here
On our planet floor,
Any broad alarm of their hastening doom
Is lost in the gloom of dust and ages.
—Maya Angelou, *On the Pulse of Morning* (1993)

CONTENTS

FIGURES AND TABLES

Figures

Tables

PROLOGUE

A day of trudging up one and then down another ravine carved into the Cretaceous sands was taking its toll. My search for fossils had been fruitless; my mind was beginning to wander. Here on the western slope of the Colorado Rockies I wasn't quite high enough to see the valley of the White River or the town of Rangely to the north. I could, however, just detect through the afternoon haze the top of the solid-walled Uintas poised above Vernal, Utah, and Dinosaur National Monument some sixty miles to the northwest. Sunny afternoons always brought this summer haze—some portion of which was natural, some from forest fires, but some certainly born of human activity. At least in the mornings from our camp, about five miles north and two hundred feet higher, I could usually find the snowy peak named after the famed nineteenth-century paleontologist Othniel Charles Marsh.

Although not exceedingly hot, it was a very dry day. This had been a bad fire season, and I could smell burnt juniper on the wind. As I sat down in the shade of a small piñon pine to channel my thoughts into something more useful for field work than the image of a cold beer, I spotted a small shiny object. To this vertebrate paleontologist it looked suspiciously like a tooth of a carnivorous dinosaur, a theropod. I picked it up, turned it over in my hand, and in an instant knew this was a keeper. The inch-long, black tooth was not large by theropod standards, but it was quite well preserved. The tiny, steak knifelike serrations along both front and back of the slightly curved tooth were readily apparent to my touch, if only just visible. I forgot about cold beer, at least for the moment, and switched into a search mode, honed over two decades of fieldwork.

I had found the tooth midway up a small ridge about ten feet high and fifty feet long. A search up the small rill that produced the tooth also turned up a few small fragments of a turtle shell; higher up I began to encounter variously sized slivers of dinosaur bone. As I neared the crest of the sandstone-capped ridge, the bone fragments became fewer and fewer and finally vanished. Looking from side to side, I could see the bone slivers concentrated in small rills dissecting the low sandstone ridge. By themselves, such bone fragments meant little. Just a day before, we named a similar site "Dinomunge" because of all the scraps and fragments of dinosaur bone strewn along the hill. Still, we needed to check every telltale sign of bone just in case something more substantial lay within. So I sat down and began to scratch just at the horizon that contained the highest bone fragments visible at the surface. Several test holes in this soft sandstone revealed nothing except more bone slivers, but on what I recall as the third try, I struck bone—substantial bone. I cannot explain scientifically why I knew I had hit fossilized bone rather than just more sandstone; I simply knew.

After just an hour of chipping away at the progressively harder sandstone, I had uncovered several long, flat bones that looked rather like the pelvis of a duck-billed dinosaur. Because I do not work on the systematics of dinosaurs, but rather the tiny mammals that scurried at their feet, I could not be absolutely sure of this field identification. Years of experience digging up pieces of these creatures gave me a good deal of confidence that I was correct. The quality of bone, unfortunately, was not very good. Each bone was encased in a fine mesh of tiny root hairs. Nevertheless, every reasonably sized dinosaur fragment from this Upper Cretaceous Mesaverde Formation was important, if only because vertebrate fossils were rarities here. If I had been working in my old stomping grounds of the Hell Creek Formation in eastern Montana, where dinosaur fragments are as common as cactus, I most certainly would have stopped digging an hour earlier.

Dinosaurs, or any fossil vertebrates for that matter, have the nasty habit of occurring where they want rather than where you need them. The Mesaverde Formation (or, more correctly on the Western Slope of Colorado, the Williams Fork Formation) would probably never rank as a paleontological paradise. Why work in an area where dinosaurs and other vertebrates are not especially common? The simple and not too surprising answer is that to fill the gaps in our knowledge of past life, we require as diverse as possible a database. Over the past hundred years our database for the last of the dinosaurs and their numerous vertebrate contemporaries has grown tremendously, probably by tenfold in the past quarter century alone.

The public, as well as the electronic and written media, are understandably attracted to the biggest, the oldest, the newest fossils, especially if they are dinosaurs. There is nothing wrong with dinosaurmania, but what is lacking in the stories one reads in the paper or a magazine, or sees on television, is that the vast knowledge that we have about past life rarely comes from the few spectacular finds, but from the tremendous accumulation of fossil data gathered by generations of researchers whose finds never become news. The search for more such unspectacular data is the central reason I was on the Western Slope of the Colorado Rockies in the summer of 1989, hacking away at this hadrosaur pelvis.

This book is about a fossil database. Specifically, it concerns the wealth of information about vertebrates from the Late Cretaceous and into the early Tertiary. It thus brackets in time one of the greatest scientific mysteries—dinosaur extinction. This account is intended for anyone interested in these extinctions. Considerable detail is here for those who wish it. If colleagues and students can derive any new insights, or if I raise some colleagues' blood pressure (just a little), so much the better.

More than a few books have been published on the extinctions at the K/T boundary. Some even use the word "dinosaur" in the title, but to my knowledge none of the various authors deal with these extinctions from the perspective of the dinosaurs and their contemporaries. By this I mean they propose theories but do not ask the next question: "What do the fossils say?" I may not answer this question to everyone's satisfaction—and I may, in fact, draw some conclusions that many would prefer not to hear, but at least I make the fossils the focus of my discussion.

You will find some speculation here, of course, but my goal has been to remain grounded in (and that sometimes means grounded by) the evidence. What do, in fact, the fossils say? I often play the role of skeptic, throwing the cold water of empirical evidence—or, more often, pointing out the lack of any such evidence—onto the hottest and sexiest theories about dinosaur extinction. Although I do, in the final chapter, propose my own speculative scenario for the end-

Cretaceous that is grounded in the fossil evidence, I believe an equally fundamental contribution is to emphasize that we just don't have enough empirical evidence to, as yet, jump on anybody's bandwagon. We must examine what the fossils say and, more often, do not say. If the fossils are silent or contradictory, we too must rein in our proclamations, while we return to the hard and unheralded work of ferreting yet another sliver of evidence out of the proverbial dry bones.

I have benefited greatly from papers and books of countless colleagues, many of whom have graciously taken the time to answer my questions, to suffer my relentless quest for the empirical evidence upon which their ideas are based, and to engage me in lively debate. Here I shall acknowledge only the few who contributed most recently to the emergence of this book. The following people offered helpful comments on either style or content, loaned me literature, or provided more current information, for which I am most thankful: Dan Archibald, Donna Archibald, Jim Archibald, Ed Belt, Annalisa Berta, Luis Chiappe, Dewitt Coffey, Jeff Eaton, Tony Ekdale, David Fastovsky, Rodney Feldmann, Lena Golovneva, Barbara Hemmingsen, Stu Hurlbert, Gerta Keller, Mary Kraus, Tom Kwak, Paula Mabee, Charles Marshall, Mark Norell, Kevin Padian, John Ruben, Anne Weil, Mark Wilson, Joy Zedler, and Bill Zinsmeister. I or Columbia University Press was fortunate in cajoling several people into reading most or all of the manuscript for this book, for which I am forever in their debt—Brent Archibald, Gloria Bader, Laurie Bryant, Bill Clemens, Peter Dodson, Richard Etheridge, Kirk Johnson, Zofia Kielan-Jaworowska, Lev Nessov, Roger Sabaddini, Bill Thomas, David Ward, and David Weishampel. I did not follow all of their advice, especially on matters of judgment, so mistakes—whether silly or substantive—are my responsibility.

Ed Lugenbeel, executive editor for Columbia University Press, was very patient over the past few years (I won't say how many) as I continued to promise that the manuscript would be forthcoming any day now. Freelance editor Connie Barlow helped to keep my abuse of the English language to a minimum. Finally, the warmest thanks to my best friend and wife, Gloria Bader, for listening over and over again to my thoughts on the importance of extinction to the evolutionary past and our collective future.

I

An Embarrassment of Riches

One of the great pleasures in vertebrate paleontology is the opportunity to travel to exotic and even not so exotic places. For me this is especially true when I finally am able to meet a colleague with whom I have communicated for a number of years. Such was the case with Lev Nessov of St. Petersburg University. During the 1970s and 1980s Lev had almost single-handedly discovered and described a host of vertebrate species from the Mesozoic and early Tertiary rocks of western Asia. Our interests and expertise thus overlapped, not only with respect to taxa but also in time. Ever since graduate school in the 1970s I have been tracking what the fossils say about when and why the dinosaurs perished. Lev Nessov had unearthed important clues from the other side of the planet.

Lev and I finally had a chance to meet in 1990 at a conference in Almaty, the capital of the Republic of Kazakhstan within the then still extant Soviet Union. A hundred paleontologists and geologists had come from all over the world to discuss "Upper Cretaceous terrestrial correlation" with their Russian-speaking colleagues. My contribution was a talk on the correlation of fossils within the Upper Cretaceous rocks of North America—that is, about how to determine the age of the rocks and the fossils they contain.

The conference was held in a spacious and well-appointed hall that contained a large, rather intricate stone mosaic of Vladimir Ilyich

Lenin. My first slide was of Edward Drinker Cope, one of the two most famous (or infamous) American paleontologists of the late nineteenth century and the first to describe Late Cretaceous mammals. The resemblance between the stoic Cope and the heroic Lenin was striking. Both had gaunt, mustachioed faces and a small goatee, although Cope did have more hair. I could not resist calling attention to their resemblance, hastening to add that I was certain the two were unrelated.

Dead silence. Had I offended my hosts or had the humor been lost in the simultaneous translation? I knew then what it means to die on stage.

For the remainder of the conference I harbored the thought that maybe Soviet paleontologists take their science without humor. That thought was, however, thoroughly dispelled during the post-conference field excursion that took the adventuresome among us to the Upper Cretaceous rocks of Kazakhstan. Times were becoming rough for Gorbachev's perestroika, but the hospitality shown by our Soviet colleagues was warm and gracious. And the humor was in no short supply.

It was less than a year later, in June 1991, that I had the great pleasure of returning the hospitality by hosting Lev Nessov on part of his journey to the United States and Canada. After a few days of capitalist decadence in the warm California sun, we set off for the even warmer sun of Utah. There we embarked on a whirlwind tour of Mesozoic and lower Tertiary rocks. Even for this well-seasoned field scientist from abroad, the stark beauty of southern Utah was overwhelming. Somewhere east of Capitol Reef National Park, Lev turned to me and said that he now knew why so much vertebrate paleontology is published in the United States. Yes, the American West has some of the greatest vertebrate-bearing rocks in the world, but it is also blessed in three other ways that are almost as important—the roads to reach (or at least approach) the outcrops, the scientists to do the work, and enough money to afford the first two.

With this combination of rocks, roads, people, and money it is no surprise that the western United States (along with western Canada) has the most thoroughly studied vertebrate fossil record for dinosaur extinction at the Cretaceous/Tertiary (K/T) boundary of about 65 mya (million years ago). More important, western North America is still the only place that has yielded a nearly continuous record of the terrestrial vertebrate fauna across this boundary. To understand why these sections in North America are so critical for studying boundary events, we need to couple a perspective of deep geological time with an awareness of just how rare it is for earth's history to be adequately recorded in the rocks.

A BRIEF HISTORY OF GEOLOGICAL TIME

The K/T boundary, like any other geological time boundary, conceptually represents an instant of time in the remote geological past. Although something written in the rocks and fossils has called our attention to this particular time, a boundary is a category imposed by humans on the rock. We called the Cretaceous into being, and we determine when it ends. Moreover, "instants" are usually not recorded in the rock and fossil record. If we are extremely lucky we may be able to demarcate a thin unit of rock and a fossil record that may reliably be said to tell us about a span of time no longer than perhaps a few thousand or tens of thousands of years. Even this is exceedingly rare. With the rarest of exceptions (such as lake varves), nowhere can we say anything about what may have come and gone in a year, a decade, a century. The best we can usually hope for is a record spanning critical intervals, such as the K/T boundary. In the immensity of geological time, this boundary is minuscule (figure 1.1).

We know with considerable certainty that the earth is more than four billion years old and that life began more than three billion years ago. It is not, however, until a little more than a half billion years ago that multicellular life—the first sea animals—seems to have burst upon the scene. This event marks the beginning of the Phanerozoic eon (figure 1.1), which literally means "visible (animal) life." The name is apropos because this is when fossils of marine creatures with hard parts begin appearing in the fossil record in abundance. The Phanerozoic eon is further divided into Paleozoic, Mesozoic, and Cenozoic eras, their prefixes meaning "old," "middle," and "new."

Although certainly influenced by the rock and fossil record, the geological time scale that has come into currency is a historical construct, not a geological truth. It is overwhelming dominated by European terms and ideas, not surprisingly because Europe is where it originated. The Eurocentric bias of the accepted way to parse geological time is very clear at the next lower level within the geological hierarchy—the periods. Many of the periods take their names from places or peoples in Europe. For example, the Cambrian period refers to the Roman name for Wales (Cambria), which is where rocks and fossils of this age were first recognized; the Permian period was named for exposures of this age in the Perm region of Russia, just west of the Ural Mountains; and the Cretaceous period is from the Latin for "chalky"

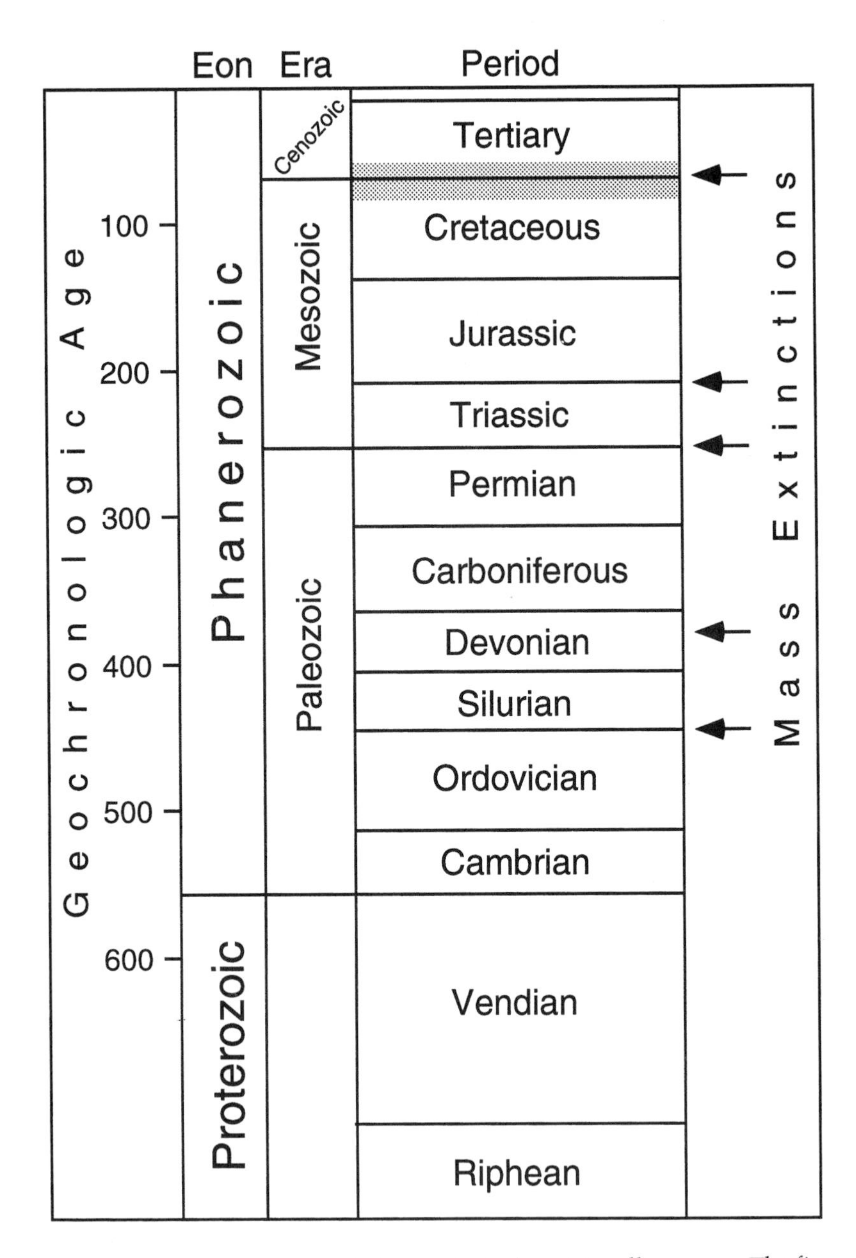

FIGURE 1.1 General geological time scale for last 600 million years. The five mass extinctions of the Phanerozoic are shown. The shaded area surrounding Cretaceous/Tertiary (K/T) boundary is discussed in this book.

(cretaceus), in reference to the extensive chalk exposures of this age in European countries abutting the English Channel.

The K/T boundary (K to distinguish Cretaceous from the C of the Carboniferous period) is the dividing line between more than just the periods for which it is named. It is also the boundary between two eras, the Mesozoic and the Cenozoic. The Tertiary, the period that constitutes the bulk of the Cenozoic, does not take its name from a place or a people. It simply means "the third" in Latin. It was so named, even before the other periods had been recognized, because it was the third part of a geological sequence recognized in northern Italy in the mid eighteenth century.

Although the names for geological periods are a historical patchwork, the concepts that these expanses of time and their names convey have been greatly extended so that they now are global in application. We are far beyond the early attempts in assessing geological time. Nevertheless, simply because we use the same names everywhere in the world does not mean that we have an equally great ability to pinpoint and correlate geological events around the globe. As you will see, even the well-studied K/T boundary interval is a prime example of how difficult it can be to correlate biological and physical events that occurred long ago. It was just this kind of correlation that drew me to the conference in Kazakhstan in 1990. Not only are the old political barriers beginning to yield but so too are the conundrums of geological correlation.

THERE'S NO PLACE LIKE HOME: THE HELL CREEK FORMATION

Although my travels, both professional and personal, have taken me to some rather exotic places, I can never completely leave behind my earlier work in the austere plains of eastern Montana. Most tourists on their way to the Rockies view eastern and central Montana as a long, monotonous haul, something to be driven through as quickly as possible. But not far off the interstate one can venture into some of the most starkly beautiful badlands in the world, easily rivaling the mountains to the west (figure 1.2).

Montana's badlands drain often-dry streams that flow into rivers with such poetic names as the Milk, the Yellowstone, and the Judith. These rivers in turn empty into one of the longest drainages in the world, that of the Missouri and lower Mississippi rivers. The badlands that have sculpted these upper reaches of the Missouri are sometimes called the Missouri Breaks; they are vastly larger than the much bet-

FIGURE 1.2 Panoramic view of the Missouri Breaks. The area shown here is near Billy Creek, looking north toward the Missouri River and Fort Peck Reservoir, Garfield County, Montana. The escarpments are mostly in the Hell Creek Formation of the uppermost Cretaceous. Photo by the author.

ter known Big Badlands of South Dakota. In the early 1930s the Missouri River was dammed in eastern Montana to form the Fort Peck Reservoir, still one of the largest in the world (figure 1.3). The newly built Fort Peck Dam graced the cover of the first issue of *Life* magazine in 1936.

The badlands of eastern and central Montana have long attracted paleontologists. In 1876, just months after Custer rode to his death and into western lore in the Battle of the Little Bighorn in south-central Montana, Edward Drinker Cope (the Lenin look-alike mentioned earlier) explored badlands along the Judith and Missouri rivers in central Montana. Only by a stroke of luck did Cope escape Custer's fate. A military post on the Missouri River was overrun and the soldiers killed just days after Cope had departed it for home (Lanham 1973).

In 1905 Henry Fairfield Osborn of the American Museum of Natural History (in New York City) named and described what has become the most famous of all dinosaurs, *Tyrannosaurus rex*. The type specimen that Osborn described came from north of Hell Creek in what is now Garfield County, Montana (figure 1.3). But Osborn was not its discoverer. Barnum Brown claims that honor. Brown prowled the badlands of eastern Montana in the first decade of this century,

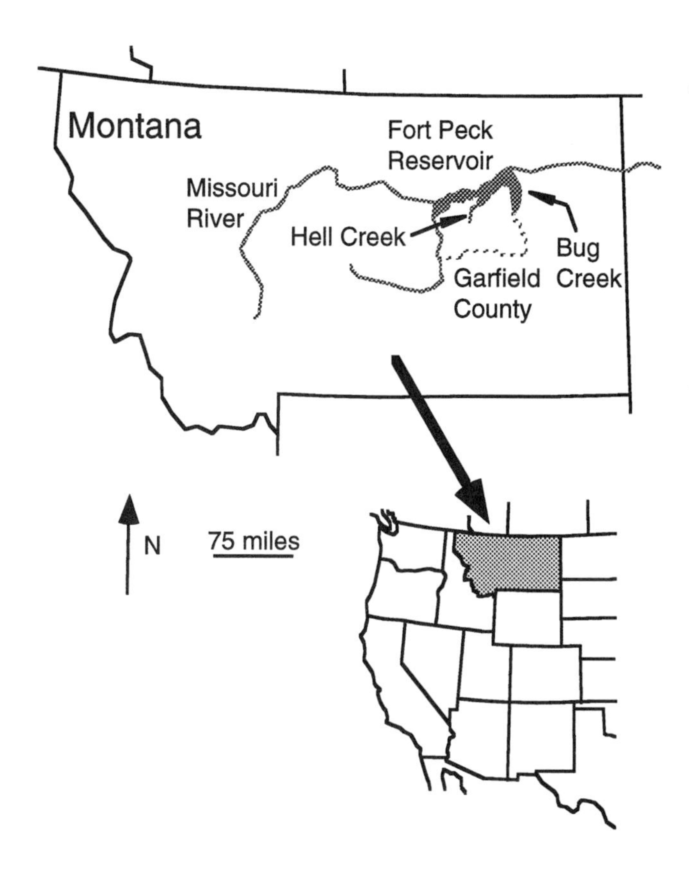

FIGURE 1.3 Hell Creek and Bug Creek areas in northeastern Montana. Both have exposed K/T boundary sections that contain vertebrate fossils.

turning up one other specimen of *T. rex*, along with a host of other Late Cretaceous dinosaurs.

The type and the second specimen originally were housed in the American Museum of Natural History. The type was moved to the Carnegie Museum in Pittsburgh during World War II, ostensibly for protection from bombing, but money also seems to have played a role. I had a chance to visit, while still a graduate student, the site that is probably the type locality of *T. rex*. In 1974 Bill Clemens, my dissertation advisor at the University of California at Berkeley and the leader of our field studies in Montana, took a number of us on a picnic to the place he had earlier identified as the probable type locality of *T. rex*, based on his examination of archival photos. There he also found bone fragments, probably belonging to the big beast itself.

But none of us had come to Montana to look for dinosaurs. Bill (then as now) was a world expert on Mesozoic mammals. He was interested not in the charismatic meat eaters but the small mammals that had scurried beneath their feet. Discoveries of small mammals made by Harley Garbani in the 1960s (and much earlier by Barnum Brown) is what brought Bill, me, and later generations of graduate students to the Hell Creek.

Today, the Hell Creek Formation of eastern Montana still ranks as one of the most productive areas for new discoveries of dinosaur fossils. It is without question the most important region for the study of vertebrates at the K/T boundary. It holds that honor, however, only because there are no contenders. The Hell Creek Formation is as yet the *only* place in the world where we have a reasonable sense of what happened to vertebrates through the K/T boundary. What makes the Hell Creek Formation so important is that it was deposited at the right geological time, some 65 million years ago when Earth's biota was beginning to experience one of its greatest upheavals.

The Hell Creek Formation is a 400- to 500-foot pile of gray and tan siltstones and sandstones (figures 1.4 and 1.5). This formation and others like it of Late Cretaceous age are expansive throughout the western states. They were laid down by rivers draining easterly into the immense but waning seaway that split North America in half throughout the Late Cretaceous. This Late Cretaceous system of sediments is similar to (though thinner than) the Mississippi river and delta system of middle Tertiary to Recent age that underlies much of the southern United States. In fact, it can be thought of as the antecedent of the Mississippi delta system. Sedimentologists call these kinds of landscapes aggrading systems, meaning that there is a net accumulation of sediment through time. Sediment was accumulating along the interior seaway during the Late Cretaceous for several reasons—the primordial Rockies were beginning to rise, the margin along the sea was subsiding, and sea level was dropping. Such an accumulation of sediment is the primary reason why dinosaurs and other vertebrates are so abundantly preserved in the Hell Creek Formation.

The Hell Creek Formation is only part of the geological story of uppermost Cretaceous and lowermost Tertiary rocks in eastern Montana. Figure 1.4 is a generalized geological section for the region. The gray shales and mudstones of the Bearpaw Shale begin the sequence. These rocks record the last of the waning seaway, including the last of the ammonites and other sea creatures in the region. Immediately above is the Fox Hills Sandstone. The 100 or so feet of this formation preserve the beaches and offshore islands that followed

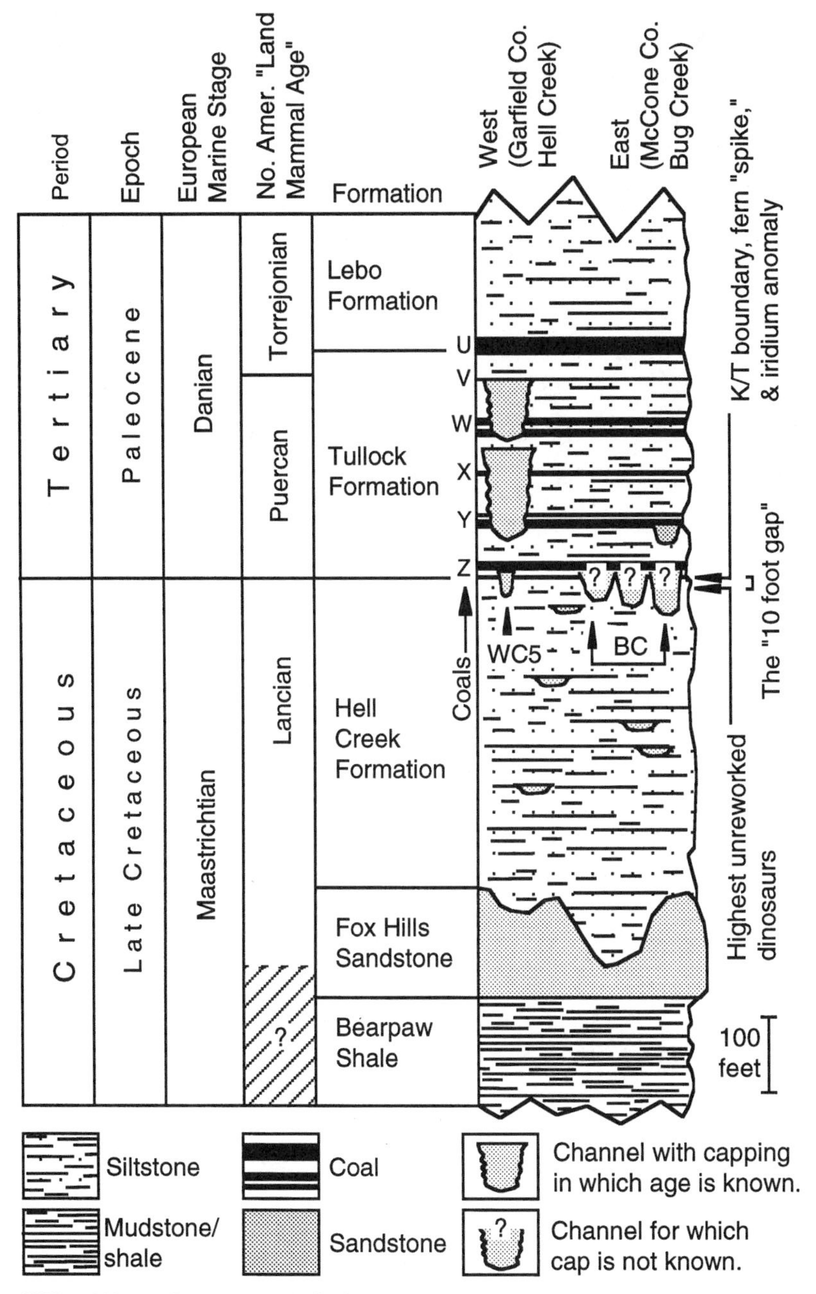

FIGURE 1.4 Generalized geological section across the K/T boundary in northeastern Montana. *Source: data from Archibald 1982, and Archibald and Lofgren 1990.*

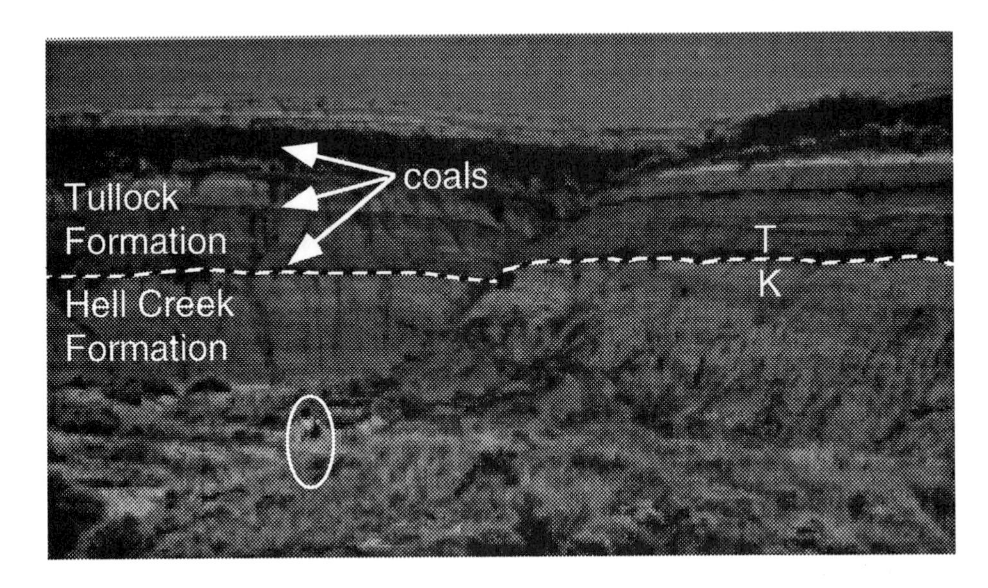

FIGURE 1.5 The K/T boundary marked by coal. In this geological section the lowest of the Z coal complex (dashed line) marks both the Hell Creek/Tullock formational contact and the K/T boundary, Garfield County, Montana. The marked coals are also part of the Z coal complex, the lowest in the Tullock Formation. The oval outlines a man, for scale. Photo by the author.

the seaway as it retreated to the east. Next above comes the Hell Creek Formation, the accumulation of rivers draining into the waning seaway. In its lower reaches the formation sometimes cuts deeply into the underlying Fox Hills. This erosion must have occurred at the beginning of Hell Creek time, as the developing streams chewed up the old beaches as they crossed and recrossed the developing low coastal plain.

The scouring of the Fox Hills by the overlying Hell Creek is not unique in the section, but it is more obvious than most. This is only because the lithologies (rock types) of the two formations are visibly different. In fact, the same kind of churning and rechurning of old floodplains and channels continued right up through the uppermost layers of the Hell Creek and on into the overlying Tullock Formation of the lowermost Tertiary. This process of reworking older rocks unfortunately also applies to fossils. Later I will discuss this nettlesome problem of reworked bones and teeth, especially of dinosaurs.

The major difference between the Hell Creek and Tullock formations comes not from a big change in the way the sediments were deposited, but in what is termed a change in base level—which includes a change in the position of the water table. The most obvious physical manifestation of this change is the first widespread appearance of coals. Coal layers start at the very base of the Tullock and reappear throughout the vertical extent of this formation (figures 1.4 and 1.5). In fact, the bottom of the lowest coal bed, which is regionally called the Z coal, separates the underlying Hell Creek from the overlying Tullock. Streams still moved back and forth across the earliest Tertiary plains, but now the sea level was beginning to once again creep upward. The rates of subsidence and mountain building had lessened; rainfall probably increased. As sedimentologists David Fastovsky and Kevin McSweeney (1987) have noted, the net result was more ponding of water, which promoted the accumulation of plant debris, eventually forming the Tullock coals.

The sedimentation that created the Hell Creek and Tullock formations was by no means continuous. The record is thus far from complete. Yet, the cycles of stream deposits throughout these formations record events that my good friend and former colleague David Schindel likened to the individual frames of a movie—in fact, a rather fragmented movie. The narrative told by the K/T boundary sections in eastern Montana is not seamlessly continuous. It does preserve, however, a series of paleontological episodes surrounding this great transition in earth history.

II

Some Myths

Misinformation abounds concerning the disappearance of dinosaurs. A partial explanation for this abundance of confusion is that many people have offered their opinions with little or no knowledge of dinosaurs. I am not sure why this is the case, but much of the ad hoc theorizing seems to come from the widespread perception that no great knowledge of these creatures, their environment, or their contemporaries is necessary to proffer a hypothesis for their demise.

Physicists would no doubt cast a jaundiced eye upon the newest theories of quantum physics published by biologists. Yet expert opinion about the demise of the dinosaurs is apparently off-limits to no one. I know of no other area of science that engenders such a plethora of published speculation. Consider: in one year two of the most prominent semipopular journals in science, *American Scientist* and *Scientific American*, published three articles purportedly about dinosaur extinction (Alvarez and Asaro 1990; Courtillot 1990; Glen 1990). All three articles included the word "dinosaur" in the title, or at least mentioned them in the first and last paragraphs. But none offered more than passing treatment (if that) of dinosaur biology or their extinction. It is even more telling that vertebrate paleontologists wrote none of these articles. No wonder there is considerable confusion in the scientific and popular literature on dinosaur extinction.

The distinction I draw between myths in this chapter and controversies in the next is meant to draw attention to statements and purported information that are clearly incorrect (insofar as we know) versus ideas that are still open to interpretation. There are three myths that are relevant to the question of dinosaur extinction. First, a very pervasive view (strengthened since 1980 by the asteroid impact theory) is that the extinction of dinosaurs was a demonstrably overnight global event. Second, there is the perception that dinosaurs form a monolithic group, not only in the size of some of their constituent species but also in their evolutionary relationships and ecology. Third, many mistakenly view dinosaurs as evolutionary failures because they are extinct. These three myths alone account for a large measure of the misunderstanding of dinosaur extinction, and accordingly I will deal with these first.

AND THEN THERE WERE NONE—BUT EVERYWHERE?

The geographically limited exposures of K/T terrestrial sequences should make us circumspect about sweeping global assertions regarding the causes and timing of dinosaur extinction. Unfortunately this is not the case, or we would not continually come upon statements in the popular press such as, "the impact theory has been buttressed by researchers who have age-dated dinosaur fossils around the world" (Perlman 1993:A5). Such explicit assertions about global records of dinosaurs are patently false, and they perpetuate one of the greatest myths of dinosaur extinction—that we have a global record of dinosaur extinction at the K/T boundary. The common "wisdom" is that dinosaurs became extinct at the same time everywhere. After all, that fact is written in the rocks, isn't it?

At some of my recent lectures on dinosaur extinction, I become rather expressive when I come to the issue of worldwide dinosaur extinction. I even thump the lectern, as a preacher possessed with the spirit. In actuality I am not given to outbursts, at least in public lectures, and I suspect the audience sees through this display of mock outrage over this myth of dinosaur extinction.

The idea that all dinosaurs exited this earth at the same time in the wink of an eye is not new. The proposition that an asteroid impact caused this very rapid decline and extinction of dinosaurs has, however, given new life to proponents of sudden death. It was proposed in 1980 in a paper by physicist Luis Alvarez, geologist Walter Alvarez (son of Luis), and nuclear chemists Frank Asaro and Helen Michel. "Extraterrestrial Cause for the Cretaceous-Tertiary Extinction" appeared in the venerable journal *Science*. The authors did not explicitly state their view of the timing of the extinctions, but the implication was quite clear that they occurred over only a few years at most. In 1987 Luis Alvarez was more explicit when he wrote, "at least one [extinction event] at the K-T horizon is basically instantaneous."

Is "instantaneous" the same in a human and a geological time scale? Even the highly educated among us generally do not comprehend the immensity of geological time. When nonscientists are told that the earth is some 4.5 billion years old, the usual responses are yawns of disinterest or expressions of disbelief. We have a hard time stepping out of our own biological and cultural time scales. We thus have a great tendency to compress events that occurred a few thousand years ago with those that occurred millions of years ago. Glaciers came and went a long time ago, and so did the dinosaurs. But to we geologists there is a vast difference between the thousands of years of glacial time and the millions of years of "dinosaur time." To most people, the pharaohs and the dinosaurs are both simply ancient. It is no wonder then that we perceive episodes of great import in the past as having taken place instantaneously. Episodes may have occurred over very long intervals of time (from a geological perspective), but we compress them into "events."

This compression of long intervals of time contributes to the common perception—even among scientists—that dinosaurs vanished from the planet overnight. We now have records of Late Cretaceous dinosaurs from every one of today's seven continents (figure 2.1). But there is a huge time and conceptual difference between Late Cretaceous and the end of the Cretaceous. Having a good supply of Late Cretaceous fossils does not mean that those fossils can tell us about the end of the Cretaceous. Indeed the record of dinosaurs near the end of the Cretaceous that has any bearing on the sudden-death dispute is almost exclusively North American. There are no global data.

The Late Cretaceous covers close to forty million years of earth history, from just under 100 mya to just about 65 mya. "Late" in this case is not something hazy. The Late Cretaceous is not simply the last few million years of the Cretaceous; it is a formal measure of geological time. The "latest" Cretaceous is more loosely circumscribed, usually meaning the last few million years of the Cretaceous. The precise interval of time that entails the Late Cretaceous is almost two-thirds the length of the so-called Age of Mammals, which is designated as beginning at the K/T boundary and which we are still in today. It is perhaps natural for the human mind to compress the forty-million-year record of the Late Cretaceous so that we can speak of global dinosaur extinction at the "end" of the Cretaceous. In truth, however, only one area in

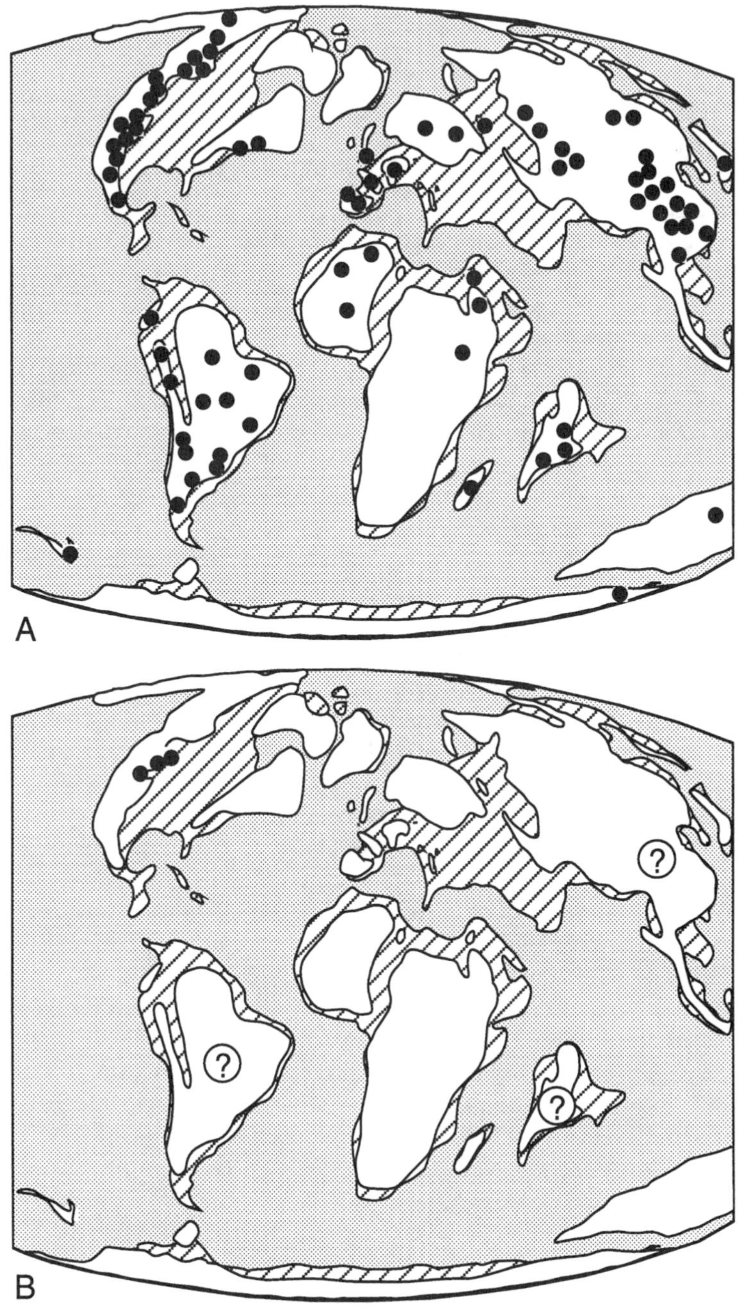

FIGURE 2.1 The paucity of dinosaur sites near the K/T boundary. (A) Paleocoastline map during the maximum marine transgression of the Late Cretaceous, some 70 mya (nonmarine areas, white; epicontinental seas, hatched); blackened circles show locations of important dinosaur sites for most of the Late Cretaceous, about 100–65 mya. (B) Locations of known important vertebrate faunas that span the K/T boundary and that include dinosaurs in the Cretaceous portion. More recently discovered, and as yet not fully explored, possible K/T boundary vertebrate faunas are indicated by circled question mark. Note the present lack of a global record of dinosaurs at the K/T boundary. *Source: occurrence data mostly from Weishampel 1990; map after Smith et al. 1994.*

the world contains fossils of dinosaurs close enough to the K/T boundary to be of any use in addressing, if not settling, the question of dinosaur extinction. This area is the Western Interior of North America, along what was then the eastern coast of a great inland sea.

Peter Dodson, a dinosaur paleontologist at the University of Pennsylvania, has shown just how skewed this record is (Dodson 1991). He limited his analysis to the last stage of the Cretaceous, known as the Maastrichtian. This stage (or age), like many in use in the geological time scale, was recognized and named in the last century, based upon a marine sequence in Europe. The Maastrichtian is about eight million years in duration, from 73 to 65 mya (Eaton 1987; Harland et al. 1989; Swisher et al. 1993). Clearly, in order to say anything about a global pattern of dinosaur extinction, one needs a global record of dinosaurs very close to the end of the Cretaceous. What Dodson found was that of the hundred Maastrichtian dinosaur localities only twenty-six are from the late Maastrichtian (near the boundary). Of these twenty-six, all but six are restricted to the Western Interior of North America. This means that more than 75% of all the information we have about dinosaur localities leading up to their extinction is from just one region on the globe. Worse, the remaining 25% of the localities are not well-known dinosaur faunas, so it is far more accurate to say that just about all of our knowledge of dinosaur extinction comes from the Western Interior (compare figure 2.1A and B). A single-region database cannot answer a global question.

In the same study, Dodson (1991) was able to show not only that the record of the last dinosaurs is very heavily biased toward North American localities, he also detected a bias in the overrepresentation of North American species and specimens. For the whole of the Maastrichtian, Dodson recorded seventy-three genera. Of these, only twenty are known from the late Maastrichtian, and fourteen of the twenty are found only in North America. These fourteen genera have yet to be discovered anywhere else. Dodson was able to identify four hundred specimens of Maastrichtian dinosaurs, but of these only a hundred are from late in this age; 95% of those hundred are from North America.

The four hundred specimens were not simply scraps. Dodson counted articulated specimens. Only articulated skeletal fragments clearly signify original deposition, rather than erosion and reburial at perhaps a much later time. This means that for each of the four hundred specimens at least several bones were found in very close association, often in the same positions they occupied when the animal was alive.

In the next few years a better global record of late Maastrichtian dinosaurs may possibly emerge. Especially promising are new finds in several sedimentary basins in China and localities in central South America. Several different groups of researchers working in southeastern China have recently reported that dinosaur skeletons, eggs, or eggshells, occur *above* what they have identified as the K/T boundary (Rigby et al. 1993; Stets et al. 1995). Articulated skeletons and eggs would not be expected to have survived an episode of erosion and reburial. The argument that dinosaurs survived the K/T boundary in China cannot thus be outright dismissed, but I believe these arguments are based on some rather circuitous reasoning about the placement of the K/T boundary.

No mammals have been reported in direct association with the highest-occurring, and thus last, dinosaurs in China. Rather, mammals occur tens of feet above the highest dinosaurs. Rigby and colleagues (1993) contend that these mammals are more advanced than the earliest Tertiary taxa found elsewhere in the world, such as mammals of middle Paleocene age in North America. The argument goes that because so few feet of sediment separate the dinosaurs from the supposedly middle Paleocene mammals, the underlying dinosaurs must be early Paleocene in age. Again, this is possible, but it makes major untested assumptions about the beginning of the Age of the Mammals and about how fast sediment was deposited between the highest dinosaurs and the lowest mammals.

In the same regions of China that yielded the "Tertiary" dinosaur eggs, Stets and colleagues (1995) detected no obvious, sharp break in the flora at the K/T boundary. Rather, the plants change gradually through the K/T transition. The data are thus provocative, and the region deserves further work, but until the K/T boundary can be more convincingly established in China, we are left with more questions than answers.

In 1988 Larry Marshall and Christian de Muizon reported on a fascinating new mammal fauna from Bolivia, which they regarded as Late Cretaceous in age. They arrived at this age assessment in large part because dinosaur tracks occurred only twelve miles away in what seemed to be the same rock unit. If there were dinosaurs, this must be the Cretaceous, or so the argument goes. The problem with this reasoning is that the mammals more closely resemble early Tertiary than Cretaceous mammals from North America. Further, twelve miles is still twelve miles. One cannot be sure that the dinosaur tracks and fossil mammals correlate in time (Archibald 1989a). The South American sites are very important, but so far they tell us little about latest Cretaceous dinosaurs.

Until we vertebrate paleontologists are able to assemble a record of dinosaur extinction that is at least passingly global, nobody is justified to argue (or worse, to assume) anything about the temporal qualities of the dinosaur extinction. You cannot say it was instantaneous. You cannot say it was gradual. You cannot even say that it was total. All you can say is that something of global proportions happened to the terrestrial biota during the K/T transition, and we are literally only beginning to scratch the surface.

DINOSAURS ARE DEAD; LONG LIVE DINOSAURIA!

One hundred fifty years ago, fragmentary remains of odd creatures were being discovered at sites throughout southern England. Richard Owen (who would later become an antievolutionary nemesis of Charles Darwin) classified these creatures as Dinosauria, a name he coined in 1842 to connote "terrible lizards." Although Owen held what we would consider as peculiar ideas regarding the origin of new species—somewhat a cross of special creation and evolution ("ordained continuous becoming")—he nonetheless was a premier comparative anatomist (Desmond and Moore 1991). He was able to take his extensive knowledge of the comparative anatomy of living vertebrates and apply similar techniques to fossil vertebrates. Owen realized that his dinosaurs were not simply lizards with a glandular problem, but a distinct evolutionary lineage.

In 1850 phantasmagoric reconstructions of dinosaurs such as *Megalosaurus* and *Iguanodon*, along with other extinct creatures, were unveiled at the great exposition in Hyde Park, London. The reconstructions are still in a good state of preservation in a quaint setting in Sydenham Park (Crystal Palace), London. Looking at the reconstruction of *Megalosaurus* in particular, it is not difficult to imagine how, from the day of their christening, dinosaurs were perceived as "terrible lizards." The reconstruction of *Megalosaurus* is more wrong than right, but the perception of a terrible lizard persists even when this animal is presented more accurately.

The persistence of the notion that dinosaurs are best regarded as monstrous lizards has played some role in obscuring the true relationships of dinosaurs among living animals. This might at first blush appear a rather incongruous statement. Dinosaurs are extinct, so how can we speak of them having relatives among living animals? The simple answer is that all extinct animals, plants, fungi, protists, and bacteria (as well as the other forms that we can no longer force into these five taxonomic kingdoms) have relatives among living organisms. If this were not the case, the single major premise of biology, that all living and extinct organisms share a common ancestor through descent with modification, would be wrong.

What is more important, however, is that, as every school child now knows (although their parents may not), the nearest living relatives of dinosaurs are birds. More correctly, bird ancestry is within Dinosauria, and thus in a very real evolutionary sense, birds are dinosaurs. This is the only unambiguous and biologically meaningful way to view the evolutionary heritage of birds. There are other, but less closely related, living relatives of dinosaurs. Traveling back along the lineage that includes birds and dinosaurs, depicted in the cladogram in figure 2.2, we find the next closest group of living relatives to be the crocodilians. Together, birds, dinosaurs, crocodilians, and several extinct groups entail the Archosauria, or "ruling reptiles."

Cladograms depict hypotheses of relationships among the groups in

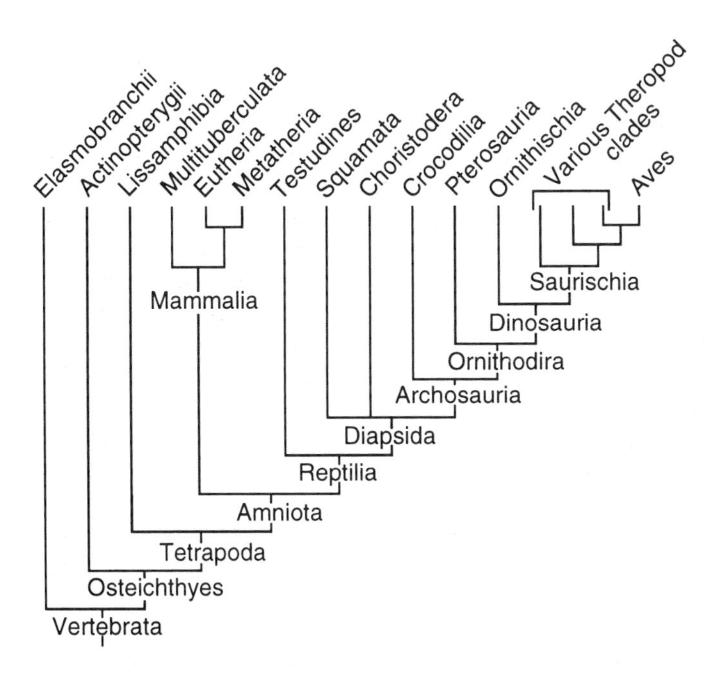

FIGURE 2.2 Phylogenetic relationships of some of the vertebrates of the Late Cretaceous. A cladogram of the major monophyletic groups represented in the latest Cretaceous Hell Creek fauna from eastern Montana. Note that in this hypothesis of relationship, birds are dinosaurs—indeed *Tyrannosaurus* (a saurischian) is closer kin to a canary than it is to *Triceratops* (an ornithischian).

question. The relationships are of a very specific kind—recentness of common ancestry. When two groups are more closely related than either is to a third, they share a more recent common ancestor. This provides the directionality in the cladogram. Recentness of ancestry is always recognized going up individual clades or lineages, from relatively older to relatively younger evolutionary splits between the lineages.

Notice in figure 2.2 that, according to the most current research, birds share a more recent common ancestry with a specific lineage of saurischians than they share with the ornithischians (e.g., Gauthier 1986). The names themselves are misleading, alas. Saurischian means "reptile-hipped" and includes such dinosaurs as *Tyrannosaurus* and *Apatosaurus*. The ornithischians are the so-called bird-hipped dinosaurs, such as *Triceratops* and *Anatosaurus*. More specifically, birds are a kind of theropod dinosaur. Theropoda, which has the confusing etymology of "beast foot," includes not only birds but all the carnivorous dinosaurs, from the huge *T. rex* to the chicken-size *Compsognathus* of the Late Jurassic. The other main line of saurischians, besides the theropods, includes the ponderous sauropods (Sauropoda means "reptile footed"). Sauropods are not shown in figure 2.2 as they are not represented in the Hell Creek Formation. Because theropods and saurischians are dinosaurs, birds are also dinosaurs. This is a deceptively simple concept that at first seems to emanate directly from Darwin's idea of descent with modification. Yet such is not the case.

Consider, for example, the living amniote vertebrates—mammals, turtles, squamates (lizards and snakes), crocodilians, and birds. As depicted in figure 2.2, amniotes share a more recent common ancestor that distinguishes them from other living vertebrates, such as lissamphibians or bony fishes (Actinopterygii). One of the strongest pieces of evidence for this particular hypothesis (any argument of relationship is always a hypothesis, although this scheme is the majority view) is a set of structures associated with the embryos of these creatures. One such commonly known structure is the yolk sac (figure 2.3), found not just in amniotes but in all vertebrates. Amniote embryos have not only a yolk sac but also an amnion, allantois, and chorion. The functions of these structures vary from one amniote to another, but some common patterns occur. The amnion, which gives the group its name, is a fluid-filled membrane surrounding the embryo. It cushions the embryo from injury. The allantois is the membrane holding metabolic wastes during embryonic development; in conjunction with the chorion, it is the site of gas exchange and nutrition in some amniote embryos. Together all three structures are the afterbirth and umbilicus of a newborn of most mammals, including humans. This unique combination of all three

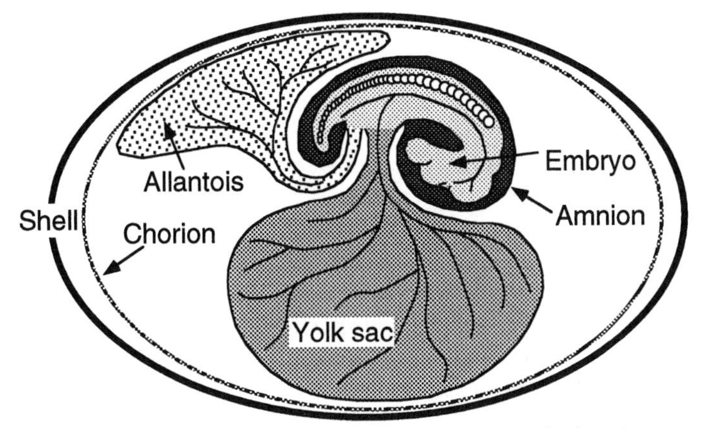

FIGURE 2.3 Amniotic egg. This schematic drawing shows the key features of an amniotic egg, such as laid by a dinosaur (including that of a bird).

extraembryonic structures is strong evidence for a common ancestor for all amniote vertebrates. Because no other vertebrate possesses even one of these three structures, the hypothesis that all amniotes evolved from a common ancestor is exceedingly strong.

The case for a single ancestry (monophyly) for amniotes does not seem to pertain when we look at relationships among the three traditionally recognized groups within Amniota—Mammalia, Aves (birds), and Reptilia. If we wish the classification of these three classes to reflect evolutionary history in the same sense that we recognize amniotes as sharing a common ancestor separate from that of all other vertebrates, then the classification would have to meet two tests. First, members of each of the three classes would have to share a common ancestor with all other members of that class. Second, that common ancestor could *not* be claimed as ancestral by any other amniote taxon. For two of the three classes, Mammalia and Aves, we can find characters, just as we did for Amniota, that strongly argue for a common, monophyletic heritage. For example, all mammals and only mammals have hair (or did) and suckle their young. All birds and only birds have feathers and a forelimb modified into a wing, whether or not they fly. These complexes of characters argue very strongly that all living mammals share a common ancestry with one another—an ancestry, moreover, that is not shared with other amniotes. The same applies for birds. When we turn to reptiles, problems arise. What characters are unique to all reptiles? What characters are possessed by all reptiles and by nothing but reptiles? The answer is none.

This conclusion certainly seems at odds with what most of us learned in school about reptiles. Reptiles are scaly, cold-blooded, egg-laying creatures. This statement is perfectly correct, but these traits are very ancient evolutionary acquisitions, simply retained by what we traditionally refer to as reptiles. These characters actually speak more to the early history of all amniotes, not just reptiles. The problem becomes more obvious when we examine these three characters.

All reptiles possess scales, but mammals and birds may also possess them. Everyone has seen the scaly tail of a rat—if not in the flesh, then in photographs or in grade-B horror films. Next time you are close enough, examine the legs of a bird. They are also covered with scales. Probably the best example is the leg of a penguin. Going down the legs, feathers merge imperceptibly into scales. This is also true for the front of a penguin's wing. Because of a penguin's need to "fly" through the water, the leading edge of its wing has tiny scalelike feathers (really just scales) that become slightly larger going back toward the trailing edge, eventually looking like small feathers. Living penguins are strong support for the argument that feathers are evolutionary derivatives of scales. Scales go back even further in the evolutionary record, as fish have them. Some modifications have, of course, occurred between the scales of fish and those of reptiles. Especially noticeable is the general reduction, or outright loss, of the bony portion of the scale among reptiles and the acquisition of a surface layer impervious to water.

So-called cold-bloodedness in reptiles is also a retention from fish and lissamphibians. The lower rate of metabolism that this term implies is retained by the majority of vertebrates. Cold bloodedness is not unique to reptiles. The so-called warm-bloodedness that distinguishes higher constant metabolic levels seems to have evolved separately in mammals and birds. Warm-bloodedness was invented more than once.

Egg laying is just as ancient as is cold-bloodedness. Among the living amniotes, not just reptiles, but all birds and even a few species of mammals lay eggs. Even though most people have heard of the duck-billed platypus and the echidna of Australia and New Guinea, many fewer make the connection that egg laying among these mammals is an ancient holdover. Even the assessment that reptiles are egg layers must be tempered by the fact that some female lizards and snakes incubate their eggs within their body, thus giving birth rather than laying eggs.

These biological facts leave us in the evolutionarily untenable and taxonomically uncomfortable position of recognizing a group of living amniotes, Reptilia, that really does not exist in any biologically meaningful way, other than if we wish to group animals on the basis of what they do not have—such as feathers, hair, etc. (The principle of absence is, by the way, precisely the logic that binds the multifarious members of another convenient aggregate: invertebrates.) This apparent paradox of the Reptilia has an easy solution, however. We must attempt to find unique characters that unite some or all reptiles with either birds or mammals. In process, the taxonomic convention of Reptilia must be modified. Jacques Gauthier (1986) performed such an analysis in considerable detail. The most interesting finding, which is quite unsettling for any one who does not automatically think in evolutionary terms, is that the nearest living relative of birds are crocodilians. More precisely, birds and crocodilians share a more recent common ancestor than do birds with any other living species. Granted, this common ancestor is quite ancient. A good guess has the Eve of this lineage walking about on all fours between 240 to 220 million years ago, in the Late Triassic. This ancestor of living crocodiles and birds was near the base of the archosaur radiation (see figure 2.2).

As we travel back in time (or down the cladogram) from the living archosaurs, we find that the extant group most closely related to birds and crocodiles is the squamates (snakes and lizards), followed by turtles, and finally mammals. If we wish to apply names that clearly and unambiguously reflect these evolutionary relationships, yet cause the fewest violations to previous usage, both Mammalia and Aves are acceptable, but Reptilia (if used at all) requires expansion to include Aves. For living amniotes we now have groups that truly reflect our best estimate of evolutionary history. The two major lineages are Mammalia and Reptilia, and you can find them on figure 2.2. Reptilia is further divisible into Testudines (turtles), Squamata (lizards and snakes), and Archosauria (crocodilians and birds).

Mammalia includes at least three major lineages, two of which are extinct. One of the extinct lineages is Multituberculata, as shown in figure 2.2. The one extant lineage includes Prototheria (platypus and echidna—not shown in the cladogram), Metatheria (marsupials), and Eutheria (placental mammals, such as us). Although my main group of interest, Mammalia, has a number of characters indicating that the relationships in the cladogram are indeed correct, I will resist the urge for further comment and return to the group of interest, Reptilia.

Although the older usage of Reptilia that excludes birds is not an evolutionarily meaningful group, I have not yet discussed characters that specifically link birds and crocodiles to the exclusion of other reptiles. In various analyses (Gauthier 1986; Benton and Clark 1988) almost forty characters drawn just from the skeleton, skull, and teeth unite crocodiles and birds with other extinct lineages within Archosauria to the exclusion of other reptiles. Other characters also support the crocodile and bird relationship among living reptiles. For example, among Reptilia (including birds), only crocodiles and birds show any parental care beyond simple nest protection seen in some snakes; both have a four-chambered heart (although a small aperture occurs between the atria, or upper chambers, in the crocodilian heart); both swallow stones (gravel for birds) to aid digestion or as ballast. The stone-swallowing character is the weakest of the lot, as it is also seen in other reptiles. Together, these characters offer a very strong case for linking birds and crocodilians to the exclusion of other living reptiles.

In the cladogram in figure 2.2, the two great lineages of dinosaurs, Ornithischia and Saurischia, are in the upper right. Birds (Aves) lie within Saurischia. This means that *Tyrannosaurus*, a saurischian, shares a more recent ancestor with any living bird, say a hummingbird, than it does with *Triceratops*, an ornithischian. One particular family of saurischians, the dromaeosaurids (including the velociraptors of *Jurassic Park* fame), is the closest to birds. Although Jacques Gauthier (1986) and Thomas Holtz (1994) differed in their interpretations of theropod relationships, both recognize a number of unambiguous characters that clearly place birds deeply within the Saurischia. Pulling back to include the Ornithischia, we see that birds are, in the truest evolutionary sense of the word, dinosaurs. Even further back we see that birds are archosaurs, and finally they belong to the great amniote clade we recognize as Reptilia.

Dissenters raise three quite different objections concerning this very strongly supported hypothesis of relationship. First, some argue that the nearest ancestor of birds is not within Dinosauria, but rather among some as yet undefined group of early archosaurs, in older usage termed *thecodonts*. Although it imparts no evolutionarily meaningful information, this term sometimes still appears in the literature (see discussion in Gauthier 1986). A variant of this objection is made by those who suggest that birds have a much closer tie to crocodilians, although the main proponent of this view, Alick Walker, has abandoned this position in favor of dinosaur ancestry (Gauthier 1986).

Second, those who fail to accept that the only unambiguous method of grouping organisms is by their recency of common ancestry sometimes raise a second objection. Birds are so different from traditionally recognized reptiles (crocodiles, squamates, turtles, and dinosaurs) that they require a grouping sensitive to these differences. Proponents of this view wish to distinguish Aves from Reptilia. But such a distinction is evolutionarily meaningless if the group Reptilia is framed to exclude Aves. Whatever battles are fought over naming, the bottom line is that there is no evolutionarily meaningful group that includes extant reptiles and extinct dinosaurs to the exclusion of birds.

Third and finally, some workers object to including birds within an expanded Reptilia and Dinosauria, even if they accept the very robust hypothesis of the dinosaur-bird relationship as true. This third objection comes from those who cannot entertain the notion that in a very real sense dinosaurs survived the K/T boundary and are with us today. This is especially true among workers advocating extraterrestrial catastrophist explanations for K/T boundary extinctions. For example, Raup states (1991 p. 7) that "the bird lineage split off millions of years before the dinosaurs died out in the mass extinction that ended the Cretaceous period. Cretaceous dinosaurs died without issue!" Raup has put himself in a bind here, because all evidence argues that Cretaceous birds are dinosaurs. Thus dinosaurs did not become extinct.

A simple, yet plausible, hypothetical scenario will show the fallacy of not using evolutionarily meaningful taxa. Imagine some time in a future millennium when celestial explorers visit Earth. They find that the mammalian biota consists of a few opportunists such as opossums and rats, a few insectivores such as shrews and moles, and many species of insectivorous bats. Just as we traditionally place scaly, quadrupedal, "cold-blooded" turtles, squamates (lizards and snakes), and crocodilians in Reptilia, while we sequester feathered, flying birds in Aves (figure 2.4A), these celestial explorers place the superficially similar quadrupedal opossums, rats, and insectivores in "Quadrupedia," while they accord the flying, grotesquely faced bats their own group, "Volantia" (figure 2.4B). It is only much later that the zoologists and paleontologists arrive to find that there was once a much richer biota, but catastrophic habitat destruction, overhunting, and most of all, human overpopulation, destroyed the biota. The bats flourished because the human chemical war on insect pests had not only failed but had backfired. Just as we now know that birds form a monophyletic clade with living crocodilians and the other extinct dinosaurs (figure 2.4C), our celestial paleontologists realize that rodents, insectivores, and bats form a monophyletic clade (the Eutheria), to the exclusion of such marsupials as the opossums (figure 2.4D).

Reframing this hypothetical case slightly differently, if the celestial explorers had followed the same logic used by Raup in his statement about birds, they would accord bats their own group (Volantia) separate from mammals because "the bat lineage split off millions of years before the mammals died out in the mass extinction that ended the Human Epoch. Human Epoch mammals died without issue!"

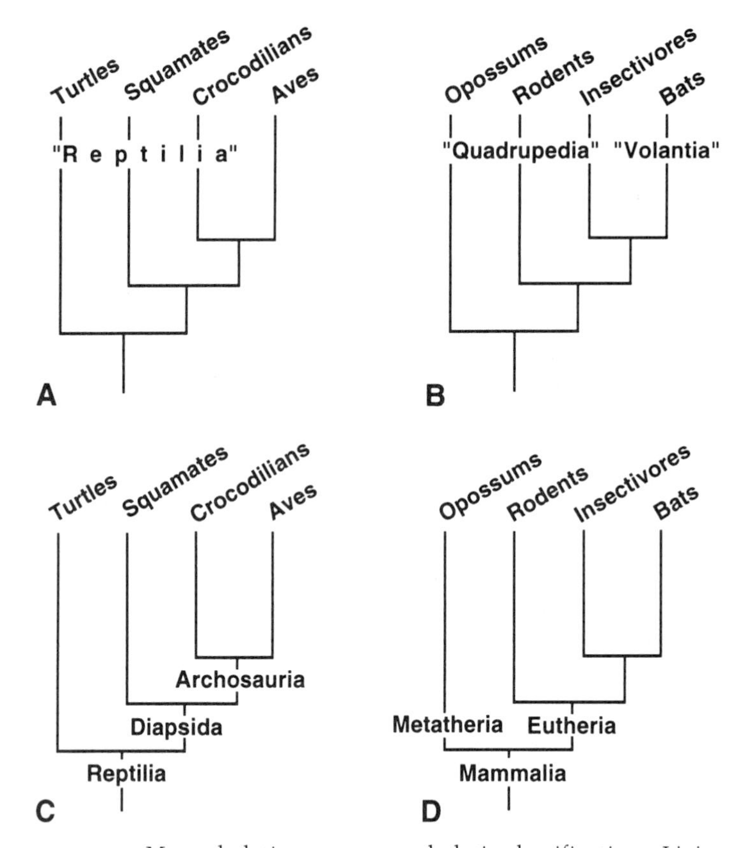

FIGURE 2.4 Monophyletic v. nonmonophyletic classifications. Living reptiles (left cladograms) and mammals (right cladograms) are classified according to nonmonophyletic (top) or monophyletic (bottom) principals. Monophyletic principles guide our modern, evolutionary classifications. For example, a "Reptilia" category (A) that joins turtles, squamates (lizards) and crocodilians because they all primitively retain scales, "cold-bloodedness," and quadrupedality, yet excludes the warm-blooded and winged birds, is incorrect; crocodilians actually are closer kin to birds (both are archosaurs) than to lizards and turtles, and the categories must reflect this kinship, as in (C). Similarly, a classification of a selected group of living mammals (B) into the four-legged"Quadrupedia" v. the flying "Volantia" would be incorrect for the same reason, as insectivores are more closely related to bats (Eutheria) than to opossums (Metatheria).

Wait, you say, bats are mammals although they are adaptively quite different from the few remaining lineages in our hypothetical future millennium. (They are even quite different from all mammals alive today.) They also made their evolutionary split from future contemporaries long before humans even appeared. The same applies in the case of dinosaurs. Birds are dinosaurs, and thus some dinosaurs survived the K/T boundary—no matter how one may try to twist the hypothesis of relationship.

The matter of proper evolutionary depiction is one thing, the problem of naming another. How, then, do we refer to the long dead creatures traditionally called dinosaurs? I think "dinosaurs" (or, more formally, Dinosauria) is fine—as long as we keep in mind that this name includes birds. For referring only to the long-dead creatures, "nonavian dinosaurs" has come into vogue, if not common usage. Unfortunately, this is a rather cumbersome, even if more correct, term. Old (even if incorrect) habits die hard, so I will continue the vernacular use of "dinosaurs" when it seems appropriate (as in the title of this book), but I will use "nonavian dinosaurs" when the distinction really matters. Nonavian dinosaurs are dead; long live Dinosauria!

GONE THE WAY OF THE DINOSAUR

The word *dinosaur* conjures different images for different people, but for most people dinosaurs as a group represent giant reptilian brutes that ran around tearing chunks out of trees or one another. Just as assuredly, we think of them as evolutionary failures. The phrase "gone the way of the dinosaur" is in many an advertiser's lexicon.

We, as humans, whose lineage split from apes four to five million years ago, or as mammals, who have dominated the large land vertebrate realm for only sixty-five million years, are mere novices compared to dinosaurs. Dinosaurs show up in the fossil record 225 million years ago, just slightly ahead of the first small mammals. While dinosaurs were the dominant large land vertebrates for the next 160 million years, the mammals remained small and their anatomy suggests that they were largely nocturnal. Only with the disappearance of the monstrous, nonavian dinosaurs did mammals explode in size and ecological diversity. A reign of 160 million years versus one of only 65 million years argues that the dinosaurs were far from failures.

The genesis and perpetuation of this myth that dinosaurs were in some way flawed owes to the tendency to equate extinction with failure. As I will discuss at length in chapter 4, extinction is the overwhelming rule and not the exception. The estimate of extinction for all species of organisms that have ever lived probably approaches 99%. Thus, with an evolutionary run of 160 million years, this is more than a passing grade and puts nonavian dinosaurs near the top of the heap, especially for vertebrates. When we add body size to their evolutionary longevity, these magnificent creatures can be considered as nothing but an evolutionary success story. If we add the birds, as we should, dinosaurs still remain an ecological and evolutionarily prominent player in today's biota. Dinosaurs have graced this planet for 225 million years. May we mammals be so fortunate!

III

Some Controversies

Controversies concerning dinosaurs abound. In this chapter I take on four that continue to influence, even if in small measure, our perceptions of dinosaur extinction. Each of the four controversies is fueled by disagreement about one or more of three fundamental questions: how many? how fast? and when? My survey takes on these four controversies in the order of least to most hotly contested.

ON THE SLIPPERY SLOPE OF DECLINE

In order to understand what occurred at the K/T boundary, we need a good fossil record as far before and after the boundary as we can measure. The reason for this is quite simple. If we wish to understand the pace and magnitude of faunal turnover precisely at the end of the Cretaceous, we need other reference faunas bracketing the boundary. Only in the Western Interior of North America do we have well-sampled vertebrate faunas bracketing the K/T boundary. *Fauna* refers to all the species found at similarly aged localities in a well-circumscribed area ranging in size from a few square feet to tens of square miles. Faunas of similar age are grouped together as ages. The ages leading up to the K/T boundary are, from oldest to youngest, Judithian, Edmontonian, and Lancian (figure 3.1). These three ages are especially well represented in the western states (figure 3.2). The ages following the boundary are, again oldest to youngest, Puercan, Torrejonian, Tiffanian, and Clarkforkian. Sometimes referred to as "land mammal ages" because they are typified by mammals, these ages can be correlated with varying degrees of success to the standard marine stages (or ages), such as the Maastrichtian.

The marine record determines the geological time scale as a matter of tradition and convenience. Marine deposits, unlike terrestrial deposits, are widespread; and the fossils within them are abundant. But what was going on in the ocean (or in one particular European stretch of the ocean, at that) may have little bearing on the faunal shifts recorded on land. Note, for example, that the boundary separating the last two marine stages of the Upper Cretaceous correlates with rocks somewhere *within* the "Edmontonian" age on land. The K/T boundary does, however, approximate the Lancian/Puercan boundary.

In chapter 5, I will deal with details of changes in the vertebrate biota through the K/T interval, as evidenced in the bones of more than a hundred species. Here I comment only on the issue of whether dinosaurs were or were not declining in taxonomic diversity through the latter part of the Cretaceous. Whether their final disappearance was

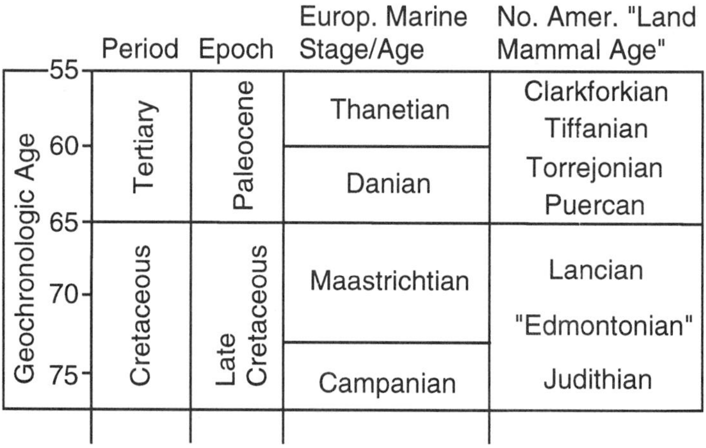

Geochronologic Age	Period	Epoch	Europ. Marine Stage/Age	No. Amer. "Land Mammal Age"
55–60	Tertiary	Paleocene	Thanetian	Clarkforkian Tiffanian
60–65			Danian	Torrejonian Puercan
65–70	Cretaceous	Late Cretaceous	Maastrichtian	Lancian "Edmontonian"
70–75			Campanian	Judithian

FIGURE 3.1 Geological time scale for epochs and ages surrounding Cretaceous/Tertiary (K/T) boundary. Boundaries between North American "Land Mammal Ages" (except for Lancian/Puercan boundary) are not yet firmly established, relative to the standard European marine stage/ages. *Source: data after Clemens et al. 1979, Archibald et al. 1987, and Swisher et al. 1993.*

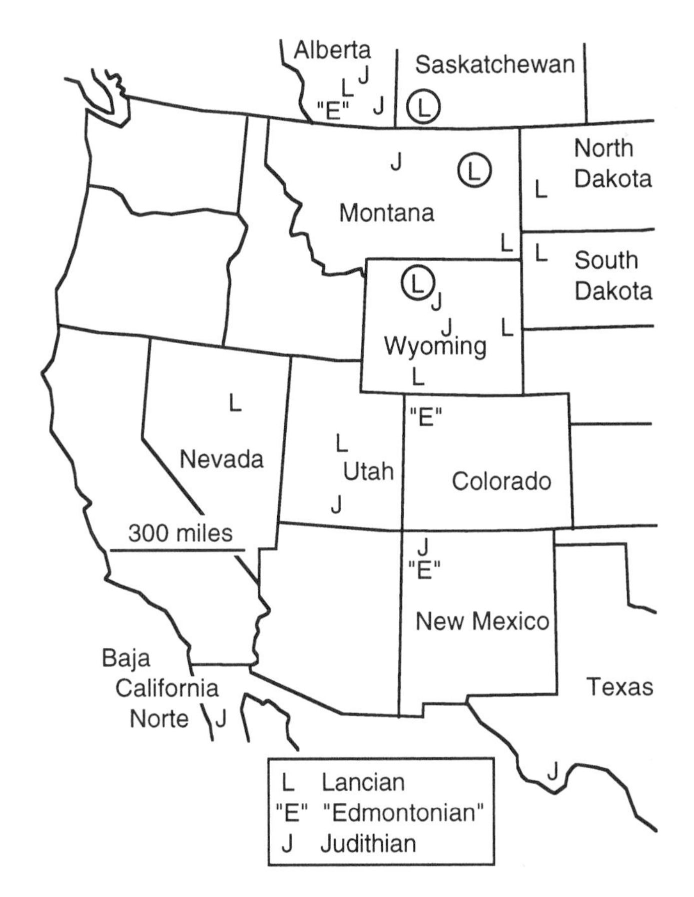

FIGURE 3.2 Vertebrate faunas of the Late Cretaceous. Shown here are sites that contain vertebrate fossils from about 75 to 65 million years ago in western North America. These are allocated to one of the last three North American "Land Mammal Ages," of the Late Cretaceous, which (from oldest to youngest) are Judithian, "Edmontonian," and Lancian. Circled Lancian sites are those that arguably include K/T boundary sections and thus that also include earliest Paleocene (Tertiary) vertebrate sites. *Source: data after Clemens et al. 1979, Lillegraven and McKenna 1986, Archibald et al. 1987, Fox 1989, and Rowe et al. 1992.*

catastrophic is altogether a different issue—perhaps the most contentious issue of all, and so I will take it up at the close of this chapter.

The vertebrate fossil record over the final ten million years of the Cretaceous in the Western Interior (comprising the Judithian through Lancian) is good. It holds the richest deposits of the best known of all dinosaurs—such charismatic creatures as *Tyrannosaurus*, *Triceratops*, and the duckbilled dinosaurs that bequeathed to posterity their numerous nests. This is probably why Peter Dodson (1991) found that latest Cretaceous dinosaur faunas are the most diverse. A high diversity in the final ten million years of the age of dinosaurs may be real; but fossil data do not automatically translate into biological truth. Taking note of the tremendously greater effort that has gone into collecting these particular dinosaur faunas, there is good reason to view with skepticism any such proclamation about a heightened diversity on the eve of destruction.

Of the three later Cretaceous vertebrate ages, only the Judithian and Lancian are well sampled (figure 3.2). We know too little of the intermediate "Edmontonian" (the quotation marks reflect our uncertainty about the distinctiveness of this age) to provide a good measure of faunal diversity in that interval of time, although I hope the work that I, my students, and several colleagues have undertaken in northwestern Colorado, which I mentioned in the prologue, will begin to fill this faunal gap.

Judithian fossils are about seventy-five million years old. Most occurrences of Judithian dinosaurs are in classic localities in the namesake Judith River Formation in southern Alberta and central Montana. Classic occurrences of Lancian dinosaur faunas are the Hell Creek Formation in eastern Montana and the namesake Lance Formation in eastern Wyoming. These two Lancian faunas are slightly younger than the Judith River faunas, spanning the last few million years of the Cretaceous. The Lancian faunas take us up to or near the K/T boundary; the three circled Lancian sites on figure 3.2 include areas that have faunas from both sides of the K/T boundary. These three regions are the best we have for studying K/T biotic turnover. The Judithian and Lancian dinosaur faunas together probably span up to ten million years and provide the best assessment of the trends in dinosaur evolution and extinction that were occurring in the waning days of the Cretaceous.

The sites I have just mentioned are not the only occurrences of Late Cretaceous dinosaur fossils in North America. They are the best-studied and the richest. Other important Late Cretaceous dinosaur faunas occur elsewhere in western Canada. Western Montana (made famous by Jack Horner) is another example. New finds on the North Slope of

Alaska are also promising. Less well known faunas have been documented in New Mexico and Texas (figure 3.2). Perhaps someday we will know these other Late Cretaceous dinosaur faunas well enough to help in analyses of the K/T boundary, especially those from the more extreme latitudes. For now, however, just those occurrences in southern Alberta, Montana, and Wyoming provide the key record of the decline of dinosaurs.

The Judith River Formation in southern Alberta has thus far been found to include thirty-three genera of dinosaurs (Weishampel 1990). Some eight or nine million years later in the Lancian, as recorded in the Hell Creek Formation of northeastern Montana, this number had shrunk to only nineteen genera (table 3.1). Most workers accept this disparity in genera between the two faunas as straightforward evidence for a decline in the number of genera of dinosaurs during the last ten or so million years of the Cretaceous (Archibald and Clemens 1982). One notable exception is Dale Russell (1984). Russell rejects the notion that the dinosaurs in the Lancian had dwindled to just nineteen genera. He argues that as many as thirty-four genera may still have been alive and well near the end of the Cretaceous, basing his case on his unpublished list of possible Lancian species. Russell, curator of dinosaurs at the Royal Ontario Museum in Ottawa, is one of the most widely respected dinosaur workers, but in this instance he has failed to sway most of the rest of us vertebrate paleontologists to his way of thinking.

Problems immediately become apparent with Russell's analysis. His report of a much higher taxonomic *estimate* of thirty-four genera for the Lancian fauna rather than nineteen *known* genera from the Hell Creek Formation is based almost exclusively on unpublished and thus unsubstantiated records. It is standard practice that if a worker uses lists of taxa, whether genera or species, in any kind of faunal analysis, there needs to be a detailed description of the fossils in question. As of yet, no list of thirty-four genera of Lancian dinosaurs from Montana or Wyoming is in the scientific literature. Someday, perhaps, such expansions of taxonomic lists will be made. But maybe not. The distinction is important, as it is my purpose here to reexamine the dinosaur extinction from the standpoint of what the fossils themselves say—not what someone thinks or hopes those fossils might eventually be shown to say. What is sound, and what is mere speculation?

Russell, in my view, also errs in using rarefaction analysis as a basis for arguing against a marked decline in genera of dinosaurs toward the end of the Cretaceous. This method estimates how many species are present in a modern biota without actually needing to count them all. Over many years of field and laboratory work, the total number of ver-

TABLE 3.1 Dinosaur Genera in the Judith River and Hell Creek Formations

Taxon	J	H
ORNITHISCHIA		
ANKYLOSAURIDAE		
Ankylosaurus		H
Euoplocephalus	J	
CERATOPSIDAE		
Anchiceratops	J	
Centrosaurus	J	
Chasmosaurus	J	
Monoclonius	J	
Styracosaurus	J	
Torosaurus		H
Triceratops		H
HADROSAURIDAE		
Anatotitan		H
Brachylophosaurus	J	
Corythosaurus	J	
Edmontosaurus		H
Gryposaurus	J	
"Kritosaurus"	J	
Lambeosaurus	J	
Parasaurolophus	J	
Prosaurolophus	J	
"HYPSILOPHODONTIDAE"		
Thescelosaurus		H
NODOSAURIDAE		
Edmontonia	J	?H
Panoplosaurus	J	
PACHYCEPHALOSAURIDAE		
Gravitholus	J	
Ornatotholus	J	
Pachycephalosaurus	J	H
Stegoceras	J	H
Stygimoloch		H
SUBTOTAL	19	10
SAURISCHIA		
CAENAGNATHIDAE		
Caenagnathus	J	
DROMAEOSAURIDAE		
Dromaeosaurus	J	H
Saurornitholestes	J	
? *Veliceraptor*		H
ELMISAURIDAE		
? *Chirostenotes*	J	?H
Elmisaurus	J	
ORNITHOMIMIDAE		
Dromicieomimus	J	
Ornithomimus	J	H
Struthiomimus	J	
SEGNOSAURIDAE		
cf. *Erlikosaurus*	J	
TROODONTIDAE		
Paronychodon		H
Troodon	J	H
TYRANNOSAURIDAE		
Aublysodon	J	H
Albertosaurus	J	H
Daspletosaurus	J	
Tyrannosaurus		H
SUBTOTAL	13	9
TOTAL	32	19

Source: Except for a few recent modifications (see table 4.1), all genera are those listed by Weishampel (1990).

Note: Judith River Formation (J), southern Alberta, is mid Campanian (Judithian); Hell Creek Formation (H), northeastern Montana is upper Maastrichtian (Lancian).

ified species in a region is established. Basically, as an area is more thoroughly worked, more and more new specimens discovered are found to belong to known species, rather than adding to the taxonomic list. This saturation of the taxonomic list is of course easier to accomplish for some kinds of species, such as mammals as compared to insects, or in areas with fewer species in general, such as temperate versus tropical areas. From repeated samplings a curve is established relating how many specimens were sampled versus how many species are expected for such a sample size. It is similar to the so-called law of diminishing returns. After you have a relatively large sample (the size can be quite variable), you can begin to estimate that you will need to sample some number of new specimens in order to recover any new species.

Using this technique developed for modern biotas, Russell generated what he argued was a rarefaction curve (Russell 1984). This curve for the latest Cretaceous is based on a little over two hundred specimens representing twenty-three genera of dinosaurs verified from the Judithian faunas of southern Alberta. Earlier in this section I noted thirty-three genera of dinosaurs from southern Alberta, based upon Weishampel 1990. Weishampel's inventory is a more recent compilation than the 1984 work by Russell. Also, Russell indicates that his count is limited to articulated specimens, which would lower the generic count. For a Lancian fauna Russell shows nine genera identified from only fifteen specimens now in the Los Angeles County Museum collection. The specimens all came from the Hell Creek Formation in Montana, which makes them about ten million years younger than the Judithian specimens.

One doesn't need a computer to see the logic of Russell's outcome. If fifteen specimens from the Hell Creek revealed nine genera, while two hundred specimens from the Judithian yielded only twenty-three, then one can surmise that the Hell Creek probably holds a lot more genera that were not represented in the museum sample. Russell indeed argued that the lower number of genera for the Hell Creek fauna is an artifact of smaller sample size.

Three problems, however, call into question his conclusions. First, it is unclear what if anything it means to compare the diversities of two dinosaur faunas based on genus as the taxonomic unit. The evolutionarily significant level is, rather, the species. This is where evolution and extinction actually occur. Traditionally, a new genus appears only because one or more new species have evolved that show characteristics distinct enough to qualify it as a new genus; a genus vanishes only when all of its constituent species have died out. In

Russell's favor, most of the genera he used in his samples were monotypic; that is, they include only one species. Genera were thus reasonable surrogates for species of dinosaurs.

A second problem with Russell's analysis is, however, more troublesome. After nearly a hundred years of being worked, there is no suggestion that the Hell Creek Formation is vastly less well sampled than is the Judith River Formation. Fifteen specimens in the Los Angeles County Museum collected from the Hell Creek Formation that Russell allotted to nine genera of dinosaurs are thus a small subset of known specimens from the Hell Creek Formation.

Third, Russell's work is not truly a rarefaction analysis. It is not generated from a known total population. Rather, it is an estimate generated by comparing two imperfectly known dinosaur faunas, and then using the better-sampled to assert that the other will have a higher number of genera when similarly sampled. The use of such rarefaction curves to postulate the "original" taxonomic diversity of one fossil fauna by using the number of taxa in another fossil fauna makes little sense. The problem is compounded by the admitted ecological differences between the two, the differences in age, and the differences in sampling techniques (Tipper 1979; Hoffman 1984).

A fourth problem emerges if, for the sake of argument, we assume that Russell (1984) is correct that diversity differences between the Judithian and Lancian dinosaur faunas are principally the result of differences in sample size. It should follow that the more poorly sampled fauna (the Hell Creek fauna) would be lacking in the rarer taxa. But this is not the case. In an earlier paper in 1967, but still the most complete of its kind, Russell tabulated that the Judithian dinosaur fauna in Alberta is dominated both in numbers of specimens and numbers of genera by two families, the Hadrosauridae (the "duckbills") and the Ceratopsidae (the "horned dinosaurs"). He found that these two families together account for about half the number of genera (12 of 23 genera) and about two-thirds of the specimens. The rarer families constitute the remaining half of the generic diversity, but only a third of the specimens. If the lower number of genera in the Hell Creek fauna was simply an artifact of smaller sample size (as argued by Russell), the rarer families should be the least well represented. This is not the case. The rarer families account for one-quarter of the specimens. The more common Hadrosauridae and Ceratopsidae account for the other three-quarters of the specimens, but represent only three or perhaps four genera. This is a drop from twelve to at most four hadrosaurids and ceratopsids. Thus, the common families, not the rare ones, declined in numbers of species in the last ten million years of the Cretaceous in these areas of the Western Interior.

The taxonomic list in table 3.1 (which is the most current available) updates that used by Russell in 1967. The general pattern just described of a decrease through time of genera within the most common families of dinosaurs still pertains and, if anything, has been reinforced by work in the past twenty-five years. This means that for the best, and really only well-sampled, latest Cretaceous dinosaur faunas in the world, the fossils themselves reveal a 40% decline in genera of dinosaurs in the waning ten million years of the Cretaceous. The dinosaurs indeed were on the slippery slope of decline long before the K/T boundary.

PALEOCENE DINOSAURS AND THE BUG CREEK PROBLEM

We humans are both fascinated and nostalgic about the last of anything—from the last Mohican, Chingachgook, in James Fenimore Cooper's 1826 novel, to the fictional Anarene, a dying Texas town in Peter Bogdanovich's 1971 film, *The Last Picture Show,* and, of course, the extinct Dodo bird. The same is true for the poor dinosaurs, even if we need to keep reminding ourselves that they are still around us as birds. Nevertheless, the really big fellows are gone.

When exactly did the nonavian dinosaurs disappear? The standard wisdom says that they all died out by the end of the Cretaceous period some sixty-five million years ago. Although few paleontologists doubt that these great beasts were already on the slippery slope of decline, fewer still believe that they made it into the Tertiary. What do the fossils say?

We can't use the presence or absence of dinosaur fossils to tell us the exact K/T boundary, as that would entail circular reasoning. I will, for the moment, assume that we can indeed establish the boundary between Cretaceous and Tertiary rocks by other means. With that assumption in place, there is then absolutely no question that dinosaur remains (the nonbird variety) occur in rocks younger than the Cretaceous. These remains, however, are probably all reworked specimens.

Reworking is nature's way of recycling inorganic and organic substances in the environment. With the exception of naturally radiated energy, very light gases, and a few stray human artifacts (space vehicles and their refuse), nothing escapes Earth. As the great tectonic plates collide, Earth's crust pushes into the depths, only to return eons later through some rift in the ocean floor. On a smaller scale, as a river meanders back and forth across its floodplain, sediments repeatedly churn and rechurn as the river creates and recreates itself. Over time, as newer sediment comes from upstream, older sediments may be reworked and redeposited with the new. If there is any subsidence of the land as this process occurs, a very thick pile of sediment may accumulate. This is what is occurring today in the delta of the Mississippi River, where a tremendous pile of sediment miles thick continues to accumulate. Although the majority of the sediments and their encased fossils are younger toward the top of the sediment pile, some mixing always occurs.

Thus at any given instant of time, a river bank is a mixture of newly deposited material and older, reworked material. If, as is often the case, organic matter combines with the mix, older and younger organic material may occur together. This is particularly true for more durable substances such as bone, shells, and especially hard enamel-encased teeth. Such mixing is the rule in rocks containing waterlain rocks and fossils.

Normally we do not attempt to detect or cannot detect this geological mixing of fossil faunas, in part because most biotic changes are occurring at a rate that is slower than the rate of accumulation and reworking of the sediment and fossils. Some fossils will surely be older than the majority they appear with—but they are not *that* much older. The remixing simply doesn't make a difference for the kinds of information we usually seek in the fossil record. When there is a dramatic reorganization of the biota in a relatively short interval, however, the problem of reworking can become very real. The reworking of dinosaurs into sediments younger than the Cretaceous is the classic case of such a dramatic reorganization.

Dinosaur teeth occur in sediments as much as ten million years younger than the K/T boundary. The Green River Formation of the early Eocene is famous for its beautiful fishes and the oldest known complete bat, *Icaronycteris*. It also yields errant dinosaur teeth. Here there is no issue, because reworking is clearly the case. As one approaches the K/T boundary from the younger side (from the top down), the situation becomes more complicated. Worm Coulee #5, a fossil locality of earliest Paleocene age a few miles west of Hell Creek in eastern Montana (see figure 1.4), has produced two worn hadrosaurid teeth. The considerable amount of wear is itself some evidence that the teeth churned and rechurned in latest Cretaceous and early Paleocene streams.

This process of reworking continues even today. The ace fossil hunter, Harley Garbani, found Worm Coulee #5 in part because of its present exposure and active erosion. In this badlands setting surely more specimens of worn, reworked dinosaur teeth will, over time, be exposed, reworked, and redeposited in what are now modern sediments. In the next round of reworking, they will mix with reworked Pleistocene bones, modern cattle, and any durable debris from ranching. So the "scientific" creationists and grade-B Hollywood movies eventually will be vindicated when the remains of humans or our culture are found with dinosaurs.

The possibility of reworking becomes considerably more difficult to judge when the geological setting is more complicated. What does it mean, for example, when instead of worn teeth you find well-preserved teeth or maybe even jaw and bone fragments? What if you find hundreds or even thousands of specimens, rather than just a few? What if the specimens you find come not from clearly datable beds that are younger than the K/T boundary, but from beds of questionable age? This is exactly the case in the Bug Creek area of eastern Montana (figure 1.3 and 1.4) and in sites in southern Saskatchewan.

The Bug Creek area is a particularly unpleasant setting to be confronted with the possibility of reworking. This is because here reworking matters. The rate of biotic change is faster than the rate of sediment accumulation. This has left us with a complicated mess that is only now being properly sorted. Although we now recognize errors of correlation made in the 1960s when these beds and their faunas were first described, the relationships of bed to fauna are the most complicated of any well-known vertebrate assemblage. To understand these complications, we must understand how the Bug Creek sequence relates to the more "normal" surrounding sediments.

As described in chapter 1, the Hell Creek Formation of Late Cretaceous age includes drab, light gray and tan sandstones and siltstones. The overlying Tullock Formation of early to mid Paleocene age is browner and more variable in color, plus it has cycles of coal seams throughout most of its vertical and lateral extent. The boundary between these formations occurs, by tradition, at the base of the lowest coal bed (which is sometimes no more than a black lignitic smudge).

The Hell Creek Formation has already furnished large amounts of skeletal remains, including complete dinosaur specimens that adorn museums around the world. The deposits from which these skeletal specimens were extracted represent both channels and the floodplains that bordered the channels. Isolated, usually shed, dinosaur teeth occur in both the channel and floodplain deposits, but more so in the channel deposits, where they were reworked and concentrated. These patterns pertain to other terrestrial vertebrates, but for some, such as mammals, skeletal material in the floodplains is so rare as to be almost nonexistent.

In the overlying Tullock Formation, bone is usually much harder to find. This was demonstrated quite convincingly in the late 1970s and into the 1980s by the field work of Howard Hutchison and Laurie Bryant. At the time, Hutchison was a collection manager and senior scientist at the Museum of Paleontology, University of California, Berkeley. Bryant was a doctoral student at the same institution. With perseverance, the two recovered very good material in some of the floodplain deposits bracketing the coal beds in the Tullock Formation. Nevertheless, the vast majority of fossil material from the Tullock has been recovered from extensive screen washing of channel deposits.

The floodplains of the Tullock are far less productive for fossils than are floodplains in the Hell Creek Formation. Moreover, no partial skeletons of dinosaurs have ever been recovered from floodplains deposits of the Tullock. The only dinosaur specimens referable to the Tullock, and hence the Tertiary in North America, are isolated teeth and bone fragments from channel deposits. This fact immediately suggests reworking from older, Cretaceous deposits. If dinosaurs had been living during the time of the deposition of the Tullock Formation, their fossils should also occur in floodplain deposits—just as they do in the Hell Creek Formation.

A weak counter to this argument is that dinosaurs had become so rare by Tullock time that they were simply too scarce to preserve as scattered remains in the Tullock floodplains. Although fossils generally are less frequent in Tullock floodplains, the careful work of Hutchison and Bryant shows no indication of dinosaurs. Thus the argument that dinosaurs were present into the Tertiary, but just very uncommon, weakens further, and reworking seems more and more probable.

Now let us return to the Bug Creek problem, which, when simplified, is a case of the mixing of reworked and original fossil material. In 1965 Robert Sloan and Leigh Van Valen reported the discovery of a series of Late Cretaceous vertebrate faunas from what they called the Bug Creek area in the western portion of McCone County, eastern Montana. These faunas included not only the traditional Late Cretaceous vertebrates such as dinosaurs, but more important, they also included new species of vertebrates that seemed to bridge the biotic gap between latest Cretaceous and earliest Tertiary vertebrate faunas.

As mammalian paleontologists, Sloan and Van Valen naturally concentrated their efforts on the mammals. They maintained that they could document a stratigraphic sequence of three localities that showed a successive evolutionary introduction of several mammalian groups. From lowest and oldest to highest and youngest these were Bug Creek Anthills, Bug Creek West, and Harbicht Hill (figure 1.4). In a series of slightly later, very thorough papers, Richard Estes (later a colleague and friend of mine at San Diego State University, now deceased) and others reported on a host of other Bug Creek vertebrates (Estes et al. 1969; Estes and Berberian 1970). Although not explicit in the original 1965 paper, Sloan and Van Valen later placed the Bug Creek faunas *between* traditional latest Cretaceous (Lancian) faunas and earliest Tertiary (Puercan) faunas. They argued that the K/T boundary lay right above the Bug Creek sequence.

At the same time, an expert collector, Harley Garbani, also was working in eastern Montana. Harley concentrated on Garfield County, which is immediately west of the Bug Creek area of McCone County. Harley, a still active paleontologist and archaeologist, has worked in eastern Montana since the 1960s, recovering spectacular dinosaur specimens for the Los Angeles County Museum. His golden touch extended to finding small vertebrate fossils as well. In particular, he found a very rich early Tertiary (Puercan) site that in 1972 caught the interest of my Ph.D. mentor, Bill Clemens, a well-known mammalian paleontologist at the University of California at Berkeley. When I joined Bill and Harley the next summer to do fieldwork in eastern Montana, the Bug Creek sequence was accepted as Late Cretaceous, although there were some rumblings to the contrary, especially from Malcolm McKenna, the Frick Curator at the American Museum of Natural History.

I accepted the conclusion that the Bug Creek localities were latest Cretaceous in age, showing a gradual or stepwise biotic turnover of mammals and some other vertebrates, leading up to the Hell Creek-Tullock formational contact and the K/T boundary. My doctoral work in the mid 1970s on localities to the west in Garfield County did not uncover any new Bug Creek-like localities. Rather, the localities even up to within about ten feet of the formational contact and the K/T boundary yielded only mammals that were typical of latest Cretaceous. This disparity led me to believe that we were dealing with two contemporary, but ecologically different, settings near the end of the Cretaceous in eastern Montana. Most sites near small streams or swamps contained only the typical latest Cretaceous mammals, along with the other vertebrates normal for latest Cretaceous faunas in the

Western Interior. A few sites, such as Bug Creek, represented much larger rivers. Fossils preserved in these less common environments included not only specimens of all the vertebrates typical of latest Cretaceous but also mammals that showed strong early Paleocene affinities.

In 1984 Smit and van der Kaars questioned the latest Cretaceous age for the Bug Creek sites. They contended that all the mammals resembling Paleocene mammals were in fact Paleocene in age. This meant that the channels themselves were Paleocene, because fossils cannot be reworked into *older* sediments. They interpreted all the abundant Cretaceous elements of the fauna as reworked. Many exchanges and interpretations have appeared since they published this argument. Although not completely settled, a Paleocene age for all the Bug Creek sites is the majority opinion.

Don Lofgren (Lofgren et al. 1990; Lofgren 1995) published the most extensive reanalysis of this problem. He worked in the area just south of Bug Creek centered in the drainage of McGuire Creek. Lofgren has found three different sorts of localities: Bug Creek-like, normal latest Cretaceous, and definite earliest Paleocene. His cautious conclusion is that all Bug Creek sites are earliest Paleocene in age and that most, if not all, dinosaur teeth and bone fragments collected from those sites are reworked from the uppermost Hell Creek Formation. The claim that nonavian dinosaurs survived into the Tertiary at least in North America is thus tenuous indeed. Paleocene dinosaurs seem very unlikely.

Lofgren's conclusion has become the consensus view of workers in the area, but definitive evidence remains elusive—and shall remain elusive. A fine skeleton in the right place could well prove the presence of Paleocene dinosaurs; but sheer absence can never prove the contrary. The weight of significant absence can, however, be taken as compelling.

As I discussed in the last chapter regarding the myth of an instantaneous global extinction of dinosaurs, there are places other than Bug Creek where Paleocene dinosaurs have been claimed—chiefly, China and South America. As I noted, however, these areas are not yet thoroughly studied, so any claim of Paleocene dinosaurs remains highly speculative and thus premature.

THE TEN-FOOT GAP

Of the four controversies dealt with in this chapter, I am most involved with the popularization (if not origination) of what has come to be known as the problem of "the ten-foot gap"—that is, the gap in the strata between the oldest reliable dinosaur fossils and the end of the Cretaceous. This furor began innocently enough from field observations made by various researchers, I among them (Archibald 1977).

As part of my doctoral fieldwork in eastern Montana in the 1970s, I expanded the mapping of the geological contact between the Hell Creek and Tullock formations begun by Bill Clemens in the area where Barnum Brown had first recognized and named the Hell Creek Formation and discovered the type specimen of *Tyrannosaurus rex*. We were also interested in trying to more accurately place the K/T boundary. Luis Alvarez and colleagues had not yet published their (1980) ideas on impacts and dinosaur extinction, so there was relatively little interest in the issue of exactly where the K/T boundary lay within the sediments. After all, until the Alvarez group proposed that a layer of iridium-rich sediment should be found at the K/T boundary worldwide, looking for a sharp boundary in both marine and terrestrial environments would have seemed as silly as looking for a signpost reading "K/T Boundary." Worse, the easiest field definition for finding the K/T boundary was no help at all from a paleontological standpoint, as the argument was to some extent circular.

This rule of thumb had been popularized by the paleobotanist Roland Brown (no relationship to Barnum Brown) in the 1950s. His advice: find the highest in-place dinosaur remains. The first laterally consistent coal above these remains, no matter how thin, marked not only the contact between the Hell Creek and the Tullock but also an approximation of the K/T boundary (figures 1.4 and 1.5). Dinosaurs had to be "in place" to ensure that there had been no reworking of specimens. Thus if bones are not at least partly embedded within the sediment when discovered, there is always a chance that man-made or natural forces carried and discarded them on the slope.

The problem of misleading fossils is more common than one might think. I recall my youngest brother, Brent, showing me a fossil fern leaf that he found along the slope of a road in eastern Pennsylvania. I no longer recall the precise identification, but even as an undergraduate geology student I was certain that no local rocks produced it. Rather, it had come from exotic materials dumped as fill during road construction. More recently the proprietor of a restaurant and bar in Rangely, Colorado, showed me a strange tooth, later identified as the milk tooth of an extinct marine mammal called a desmostylian. The nearest living mammals that resemble desmostyles are the sirenians, such as the endangered and thoroughly aquatic manatees of Florida, except the desmostyles still retained legs that allowed them to walk along the shallow sea floor. After some sleuthing among my paleontologic colleagues, it was clear the tooth must have come from a handful of sites in California that produced this kind of specimen. Yet the gentleman said that he found the tooth in a load of gravel in northwestern Colorado. Somehow it had found its way to Colorado. The owner of this tooth did have relatives in California, so he may simply have wished to trip up the college professor. (Being of a trusting nature, I prefer the first explanation.) These two stories of fossil displacement are extreme, to be sure. But in paleontological work, one must remember that vertical displacement of a fossil by even a few feet may wreak havoc with conclusions.

To return to my story of mapping the contact between the Hell Creek and the Tullock in the 1970s, I knew it was possible to find a marker for the K/T boundary that would be more precise than Barnum Brown's old rule of thumb. The greatest change in kinds of pollen and spores occurs within or at the base of lowest coal above the highest in-place dinosaurs. Although pollen and spore grains are usually not visible in samples examined in the field, they are almost always far more plentiful than dinosaur remains. The change in pollen and spores is unarguably a much more precise (repeatable) marker of the terrestrial K/T boundary than is the absence of dinosaur bones and the presence of the first coal seam. How accurately palynofloral (pollen and spore) change correlates to the K/T boundary as originally defined in European marine strata is another matter. Nevertheless, it was probably the best that we could hope for at this time.

While doing my mapping, it struck me that unreworked and in-place dinosaur remains do not occur within the Hell Creek Formation any closer to the presumed boundary (the lowest coal) than about ten feet. Equivalently aged beds to the north in southern Canada and to the south in eastern Wyoming show a similar pattern. Even though dinosaur remains are relatively rare compared to other smaller vertebrates, plants, or pollen, I reasoned that the deposits near the K/T boundary cover thousands of square miles in the northern part of the Western Interior; thus the lack of even partial dinosaur skeletons within ten feet of the boundary was unquestionably a real phenomenon. Possibly dinosaurs became extinct, at least in the northern reaches of the Western Interior, before the marked pollen change in the overlying coal. Possibly dinosaurs vanished before the K/T boundary.

Researchers became much more interested in this problem of the ten-foot gap after Alvarez and colleagues published their impact hypothesis in 1980. As I discuss in chapter 7, the marked increase of the element iridium at the K/T boundary was the strongest evidence used to argue that a bolide (a comet or large meteor) impact occurred at this time. This iridium spike was alleged to be both real and directly correlated with the mass extinction. When iridium enrichment was discovered in eastern Montana, it always occurred in association with the lowest coal, lying at the Hell Creek–Tullock formational contact. Once again this was about ten feet above the highest dinosaur remains. Various authors, me among them, noted that if this ten-foot gap is real, dinosaurs disappeared at least from eastern Montana before the hypothesized bolide had struck the Earth.

Even before the ten-foot gap became an issue, advocates of the impact theory were upset that paleontologists were not accepting their theory without objection. In a 1983 paper in the *Proceedings of the National Academy of Science*, Luis Alvarez wrote, "So my biggest surprise was that many paleontologists (including some very good friends) did not accept our ideas. This is not true of all paleontologists; some have clasped us to their bosoms and think we have a great idea" (1983:632). In the same paper, he included a long section that attempted to explain away the gap as a problem in statistical sampling. As he noted in reference to the ten-foot gap, "I really cannot conceal my amazement that some paleontologists prefer to think that dinosaurs, which had survived all sorts of severe environmental changes and flourished for 140 million years, would suddenly, and for no apparent reason, disappear from the face of the earth (to say nothing of the giant reptiles in the ocean and air) in a period measured in tens of thousands of years" (p. 639).

I think that one of the problems was that the physicist's penchant for accuracy had met the reality of the vagaries of the fossil record.

Alvarez argued that the gap between dinosaurs and the iridium-bearing coal was a sampling error caused by the relative scarcity of dinosaurian remains. One immediate problem with this explanation is that scarcity of dinosaur fossils is not a phenomenon localized in eastern Montana; it is observable at a variety of localities scattered over the northern part of the Western Interior. Thus scarcity of dinosaurs is clearly not a valid criticism.

More to the point, Alvarez had not applied proper statistical methodology—a point that seemed to escape all but Malcolm McKenna of the American Museum of Natural History. Alvarez made the unsupportable assumption that dinosaurs are randomly spaced within any given vertical section of rock. The problem is that dinosaurs in the sediment are not at all like electrons of an atom that follow some model of probability of occurrence. He needed, however, to make this assumption; without it, the pertinent calculations could not have been made.

With this assumption of random spacing, Alvarez next used figures about dinosaur placement in the uppermost Hell Creek Formation, garnered from unpublished data of Bill Clemens, to conclude that, "In the case of dinosaur fossils, the average spacing is unknown, but in Bill Clemens' table it is slightly more than 1 meter [about 3 feet]. If we took it to be exactly 1 meter, and independent of lithological factors, the analytical expression for the chance that the iridium layer appeared at least 3.4 meters [about 11 feet] above the highest fossil is $p = e^{-3.4} = 0.033$" (1983:639). By his reasoning, a gap of eleven feet between the highest dinosaur and the iridium spike (or any gap of this size), as occurs in this section, has a probability of only 3% of occurring.

As McKenna told me (personal communication, 1985), this result is exactly opposite of what Alvarez was actually attempting to show. If one is trying to forge a causal link between an asteroid impact and the highest dinosaur specimen resting eleven feet below, a probability of only 3% that an interval of this magnitude could occur by chance is indeed a small probability on which to claim causation.

This episode is a vivid example of how scientists coming from vastly different fields of expertise (in this case, physics and paleontology) can sometimes view the natural world in utterly different and incommensurate ways. In all realms of science researchers try to reduce scientific ideas to their most fundamental and universal forms; for physics this may be as simple as a mathematical equation, but the complexity of biological and geological systems very rarely affords such a luxury. In my view, there is no statistically valid way to explain away the ten-foot gap. The gap exists; until somebody finds a fossil higher up in the formation, the gap must be deemed as real—whatever its cause. The fossils themselves must have their say.

Walter Alvarez, geologist son of the senior Alvarez and coauthor of the 1980 Alvarez et al. paper, literally illustrated just how divergent these views of the natural world can be. In 1991 he published a photograph of Luis standing in Bottaccione Gorge, just outside the picturesque town of Gubbio, Italy. This is where the iridium spike had first been reported at the K/T boundary. The photo caption reads, "Geology is more complicated than physics: When physicist Luis W. Alvarez visited the K/T boundary at Gubbio, it disturbed him that the beds were dipping at 45°. He leaned over and had this picture taken with the camera tilted, so that audiences of physicists would understand the originally horizontal beds" (W. Alvarez 1991:34).

Although Luis Alvarez's statistical test fails, does this mean that one can argue that the gap represents a true disparity in time between dinosaur extinction and an asteroid impact? I think the answer is no, because on further examination, and for some as yet undetermined reason, the uppermost part (about ten feet) of the Hell Creek Formation is much less fossiliferous than is the remainder of the formation. Howard Hutchison and Laurie Bryant recognized this pattern while they were examining the lower portion of the overlying Tullock Formation in the late 1970s and early 1980s. Some possible reasons for this general paucity of fossils close to the Tullock are the leaching of bone near the formational boundary, the nondeposition of bone because of low rates of sediment accumulation, soil development that might destroy bone, or some other cause. For now all that we can argue with certainty is that this ten-foot gap between the last of the dinosaurs and the iridium enrichment in the northern part of the Western Interior is real, but that it is a gap for most vertebrates not just dinosaurs.

Newer discoveries farther to the south in the Western Interior are beginning to yield information regarding just how close dinosaurs came to K/T boundary in that region. Unlike farther north in Wyoming, Montana, and Alberta, where the presence of the last dinosaurs is argued on the basis of bones, the information for the last dinosaurs in Raton Basin of northeastern New Mexico is based on footprints. I, as well as other paleontologists, have tried to find bone, any kind of bone, in the Raton Basin, but with little luck. The area preserves beautiful plant fossils across the K/T boundary and now is beginning to yield trackways of dinosaurs.

The geologist Chuck Pillmore and a number of other earth scientist colleagues have reported finding hadrosaur and ceratopsid tracks three to six feet below the palynologically (pollen and spore) defined K/T boundary. One horizon of hadrosaur tracks is reportedly only fifteen inches below the boundary (Pillmore et al. 1994). There is no way to be sure of just how much time is represented by these fifteen inches in the south compared to that represented by the ten-foot gap farther to the north. All things being equal, however, the Raton depth probably represents a considerably shorter amount of time. The fifteen-inch position in the south is likely closer to the boundary in a temporal sense than is the ten-foot position in the north. Unfortunately, the tracks do not reveal whether dinosaur populations were in decline or doing well. What these tracks do show unequivocally is that the animals were alive at the time the tracks were produced, and thus dinosaurs in Raton Basin were alive to within fifteen inches of the K/T boundary. This is what the fossils say, but they say no more.

OUT WITH A BANG OR A WHIMPER?

The title of this fourth controversy was also the title of an essay that Bill Clemens, Leo Hickey (a paleobotanist at Yale), and I wrote for the journal *Paleobiology* in 1981. T. S. Eliot gave us the perfect way to cast the controversy that had erupted over the Alvarez theory. While I cannot speak with absolute certainty for my coauthors of this essay, as of the writing of this book I think it fair to say that Hickey supports the Alvarez impact theory while Clemens does not.

Hickey apparently had a miraculous conversion, somewhat like Saul on the road to Damascus. In reference to the idea that an asteroid might have caused global fires, he once called the Alvarez theory the "flash fryers." The last time I spoke with him, which was several years ago, he had taken on the persona of a true devotee and could not believe that all, including me, had not come to accept the Alvarez version of events at the K/T boundary. Clemens, however, seems to still have grave misgivings about the impact theory. In fact, in a recent interview, Clemens is quoted as saying that "the results of studies of patterns of survival and extinction of vertebrates *fully falsify* [my emphasis] the hypothesis that an impact caused the series of environmental catastrophes embodied in the 'Dante's Inferno' scenario" (Glen 1994:245). The Dante's Inferno to which Clemens refers is a phrase used by Walter Alvarez (1986) to characterize all the horrible events that have been suggested as corollaries of the impact. In chapter 7 I return to a discussion of Dante's Inferno and how we can test the corollaries using, as Clemens called it, "patterns of survival and extinction of vertebrates."

Nowhere are the interpretations of these events more controversial than when we turn to the issue of "how many" and "how fast" for dinosaur extinction. Did the dinosaurs go out with a whimper or a bang at the K/T boundary? As you will see, on this issue too I argue that we can't say it was a whimper; we can't say it was a bang. This is because the fossils have chosen thus far to remain as silent as the stones they have become.

I argue in this conservative vein precisely because of the first myth of dinosaur extinction discussed at the beginning of chapter 1. This myth warrants repeating. Our record of dinosaurs at or near the K/T boundary, and the vertebrate faunas that immediately follow them, are known from only one area in the world; this is the Western Interior, especially eastern Montana.

One side of the issue, most strongly defended by Robert Sloan of the University of Minnesota and Leigh Van Valen of the University of

Chicago, is that dinosaurs decrease gradually both in numbers of individuals and numbers of species as one approaches the K/T boundary in eastern Montana. Until a few years ago, I also defended this thesis. In the most recent version Sloan and his colleagues (1986) argued that the number and variety of isolated dinosaur teeth decrease as one approaches the K/T boundary in eastern Montana. The detailed defense of these measures is not published, and thus cannot be evaluated, although Laurie Bryant (1989) did not find such a decrease in her studies of the vertebrates in the Hell Creek Formation. Most important, however, the study of Sloan and Van Valen was for isolated dinosaur teeth within the Bug Creek sequence. As discussed earlier, most—if not all—of this sequence represents channeling and reworking from the overlying Paleocene Tullock Formation. Any decline would thus be an artifact of less and less reworking of fewer and fewer dinosaur teeth as one goes into younger beds.

On the other side of the controversy, Peter Sheehan, an invertebrate paleontologist at the Milwaukee Museum, and his colleagues published a study in 1991 that purports to show no decline in either the kinds of dinosaurs or numbers of individuals as one's sampling approaches the K/T boundary. The authors indicate that they were tracking the diversity of eight families of dinosaurs vertically through the uppermost Cretaceous Hell Creek Formation. They say they were testing whether they could or could not detect any change in the relative abundance of individuals or families of dinosaurs as one approaches the K/T boundary. They reasoned that if there was no change, it meant that the diversity of dinosaurs did not decrease when approaching the K/T boundary. If, however, change was detectable in the relative abundances, this would indicate a diversity change approaching the K/T boundary. The study concluded that there was no discernible change. Sheehan and colleagues surmised that this conclusion was compatible with theories of catastrophic extinction. Unfortunately, their methodology precludes them from addressing the question they wished to answer—whether there was discernible change.

According to Sheehan et al., the eight families of dinosaurs that formed the basis of their study are represented by fourteen genera. They used family-level data in their analysis rather than the fourteen genera because they felt that generic level data could be misleading and that many otherwise good fossils would have to be excluded because they were identifiable only down to the family level. This means that genera (not to mention species) could well have become extinct within the Hell Creek Formation without being detected by the authors. The distribution of genera in families used in their study shows that six of the fourteen could have become extinct without any drop in familial diversity. Thus as much as a 43% generic extinction could have taken place without the loss of a single family. This study used a taxonomic level (family) that is far too coarse to detect a decline in dinosaur diversity as one approaches the K/T boundary, if there was such a decline.

For very general statements in biology and paleontology, we can sometimes use higher categories such as families or orders to convey the broad outlines of diversity. For example, in teaching a general biology class, I might note how there are more families of placental mammals as compared to families of marsupials in order to convey the impression that placentals show a wider diversity of form (bats, blue whales, people, bears) compared to marsupials (kangaroos, opossums, phalangers). Such comparisons can often be very deceptive if one is not careful. Many people are familiar with the distinction of animals as being either invertebrates or vertebrates. This comparison strongly colors our view of animals because it is at such an inclusive or high level of categorization; yet one of the two groups in this seemingly fundamental binary has no evolutionary meaning. To drive home the point for my students, I explain that the category of vertebrates can be likened to the category of Californian. Living in California is something all Californians "possess," just as all vertebrates possess vertebrae. But finding someone to be non-Californian or an animal to be other than a vertebrate (an invertebrate) says almost nothing. Vertebrates and invertebrates are not equally meaningful and balanced categories.

What is the lesson of Californians and non-Californians for the dinosaur extinction question? The analogy warns us that presumed higher level groups may be artifacts of our modes of classification. Above the species level we can get into trouble if we use higher taxonomic levels as proxies for processes operating at the species level. We must be wary that high level groups might be serving to enforce our biases—whether those biases pertain to people or dinosaurs.

In response to my, as well as others', criticisms of their work, the basic defense of Sheehan and his colleagues was that their statistical measures were significant in showing no decline. Being of the Mark Twain school of thought (which notes that there are "three kinds of lies: lies, damned lies, and statistics"—Twain attributes this aphorism to Disraeli), I remained troubled about the claims of statistical invulnerability. I asked a colleague at San Diego State, Stu Hurlbert, to read the Sheehan et al. paper. Hurlbert is a statistically minded ecologist, so I thought he might show me the light. Instead, he confirmed my suspicions about this study. In the end, we wrote a paper together on this subject (Hurlbert and Archibald 1995). Of relevance here was Hurlbert's finding that the two statistical approaches (both are varieties of diversity indices) used by the Sheehan group were not only incapable of detecting whether there was decline or no decline in the kinds of dinosaurs. They could not even detect if there had been an increase!

I am not sanguine that we will ever confidently be able to answer whether the dinosaurs departed with a bang or a whimper. Geographically wider studies than those relied on by the Sheehan group will help, but the quality of preservation of dinosaurs will forever make these charismatic creatures poor subjects in the study of evolution and extinction. The issue of a bang versus a whimper is really the "how fast" part of the equation that I posed at the beginning of this section. The issue of "how many" is probably more approachable, or at least I will try to approach it in chapter 6. But, overall, the lesson here is a warning about overzealousness. We may desperately wish answers to our burning questions. But we must be true to what the fossils do—and do not—say. In the meantime, we should turn that drive into further commitments to return to the field and to refrain from fruitless, indeed misleading, forays with statistics and from oratorical excess.

IV

When Is Extinction Really Extinction?

We hear in the media almost daily about the loss of species and habitats around the world. Debates rage over the question of just how many species are disappearing annually or even daily. Whatever the number, virtually everyone agrees that extinctions are happening at a rate higher than in the recent and probably more distant past.

As with most people aware of this situation, I worry about what kind of world (if any) we will leave to future generations. As a paleontologist I have a further interest, as well. I study one of the allegedly greatest episodes of extinction from the dim, distant past of Earth's history—the demise of the dinosaurs. I say allegedly, because as will become apparent in this book there are many misperceptions of dinosaur extinction.

A very real problem in analyzing the fossil record is determining whether the disappearance of a fossil species from any given area is really an extinction or some other phenomenon requiring explanation. This might at first glance seem a nonproblem. Shouldn't any disappearance in the fossil record be taken as an extinction? In this chapter I will argue that a disappearance from the fossil record should not automatically be assumed an extinction. Such an assumption, moreover, can lead to severe misunderstandings of the fossil record.

A starting point for this examination is a definition of extinction. This might seem an odd and unnecessary endeavor, but as will become clear, *extinction* covers various concepts, depending upon one's definition. To begin, I define what is generally regarded as extinction as the total loss of a lineage or species and its unique combination of genetic information.

For the modern biota, we can estimate—if not actually pinpoint—when extinction, in the sense of utter and absolute loss of a species, actually occurs. For example, the last known individual of the Tasmanian Wolf, or Tasmanian Tiger, died in the Hobart Zoo in Tasmania in 1933. Unconfirmed sightings persist (Rounsevell and Smith 1982), but we can be fairly sure that this creature is extinct. This marsupial showed remarkable evolutionary convergence with the true wolf, a placental mammal found in the northern continents. It was hunted to extinction because it was regarded as a pest species preying on domestic sheep.

The extinction of the Passenger Pigeon can be dated with even more precision. A now-poignant story about the presumed invincibility of the Passenger Pigeon bears retelling here. The Passenger Pigeon was discussed at length in the first published statement espousing natural selection as a major factor in evolution. This paper was a joint venture of Charles Darwin and Alfred Russel Wallace. It was read to the Linnean Society in July 1858 by the "father of modern geology," Charles Lyell, and the well-respected botanist and director of Kew Botanical Gardens, J. D. Hooker. The reading, and publication that followed, consisted of three parts: an essay by Darwin dating from 1839 through 1844, outlining his ideas on natural selection; portions of an 1857 letter from Darwin to the American botanist, Asa Gray; and finally, the essay by Wallace written and sent to Darwin in 1858 telling of Wallace's ideas on natural selection. It is this final essay that gives us the substance of our story.

The Passenger Pigeon, according to Wallace, was a strong flier and had abundant food sources. Wallace contended that these two factors allowed this bird of eastern North America to produce just one or at most two eggs per year without any detriment to its population size. (This compares to a more common nesting pattern in other species of perhaps up to six eggs, out of which only a few survive to replenish the population.) At the time of Wallace's essay the Passenger Pigeon was a very abundant bird in the eastern United States. Some flocks of this migrator were estimated at more than two billion (an exaggeration, perhaps). Who would have thought then that the species was vulnerable? The last of the wild pigeons had been shot by 1899; in 1914 the last bird died in captivity (Smith and Maffitt 1994).

A species is usually doomed to extinction well before the last individual dies. This was certainly the case for the Tasmanian Wolf and for the Passenger Pigeon. When the population size of a species falls below a certain level, possibly as many as fifty individuals, the likelihood of recovery diminishes greatly. Only in a few cases where humans have intervened has a species been brought back from the brink. The recently instituted captive breeding program for the California Condor is designed with this goal in mind. Humans unquestionably contributed to the precipitous decline of this species, but climate changes since the Pleistocene already had begun to take a toll long before Europeans arrived on the scene. In the 1980s the last wild condors joined their captive brethren in breeding programs at the Los Angeles and San Diego zoos, with a total of forty in captivity by 1991. Captive breeding has been successful, but attempts at reintroduction have not yet been. This is because of habitat loss or degradation—a problem that casts a shadow on the futures of many such "show" species fortunate enough to attract human attention and dollars for captive breeding.

When we turn from the present and near past to the long distant past, we cannot know with certainty the year or even millennium when a species became extinct. We will never have the temporal luxury of knowing when the last trilobite foraged in the sea or the last dinosaur trod the earth. What we find in the rock record, if we are extraordinarily lucky, is the apparent disappearance of a species over a very wide geographic range within a relatively narrow stratigraphic (and hence, time) interval.

EXTINCTION AS THE RULE, NOT THE EXCEPTION

Despite the unavoidable vagueness as to the timing of prehistoric extinctions, we can be certain that the number of past extinctions is exceedingly high. Estimates that reoccur in the paleontological literature suggest that over 90% and possibly over 99% of all species that have ever existed are now extinct (e.g., Raup 1991)! Although such figures are obtained with a healthy dose of extrapolation, limits can be set that show they are not far off the mark. To demonstrate this I will show how such high percentages have been obtained using data that is easily gleaned from the paleontological literature. We only require three estimates. The first estimate is the average duration for a species. For this, tabulate the known geological time ranges for as many extinct species as possible or, as in the next paragraph, find a source that kindly has done the tabulation for you. We tend to avoid extant species in such tabulations since we do not know their "full measure." The second estimate may seem easy—the length of time the earth has supported life. Early life, however, may have been very simple for a very long time, perhaps 3 billion years. A better estimate would be the time since the origin of hard parts that could fossilize (550 mya), a point which marked an episode of explosive speciation. As I will show, either estimate will suffice in arguing that well over 90% of all species that ever lived are extinct. The last of the three estimates is the number of species alive at any given time. As with the estimate of species longevity, we can use tabulations found in the literature to suggest this number.

I calculated an average species duration of 12.4 million years by averaging all the mean species durations shown by Niles Eldredge (his table 3.1, 1989). Eldredge noted eleven different groups of organisms and found that the life spans of species ranged from a high of twenty-five million years (for a variety of microscopic marine creatures: benthic foraminiferans and marine diatoms) to a low of just a million and a half years (for mammals).

A generally accepted figure for the origin of life or at least the earliest known life is about 3.5 billion years ago. The oldest well-established fossils are preserved in the Warrawoona Formation in western Australia (Schopf and Walter 1983). The fossils are of cyanobacteria (formerly known as blue-green algae) that today build, as they did billions of years ago, mounded structures called stromatolites. Although restricted now to very salty nearshore environments, out of the way of grazing mollusks, cyanobacteria were much more widely distributed before their predators evolved.

For an estimate of the number of species alive at any given time, the number of species alive today is the place to begin. Figures vary widely for the number of extant species. In his intriguing 1992 book on biodiversity, E. O. Wilson put the number of *known* living species at 1.4 million, but he judged that the *actual* number may be anywhere from 10 million to 100 million. Whatever the actual number, it has certainly declined, as we have launched the Holocene down the slippery slope of modern mass extinction.

The fossil record indicates two major jumps in species diversity since the beginning of the Phanerozoic some 550 mya. The first occurred over the course of the Cambrian and Ordovician. The second happened near the beginning of the Jurassic, some two hundred million years ago (figure 4.1). The premier chronicler of the diversity of marine invertebrates, Jack Sepkoski, thus discerns three major evolutionary faunas with this three-stage pattern (e.g., Sepkoski and Miller 1985). Overall, there is widespread agreement that the biota is much more species rich today than it was for most of the expanse of geological time. Thus we would greatly overestimate the average number of species present at any given time if we used the (10–100 million) estimated by E. O. Wilson for today's biota. A reasonable way to proceed would be to take Wilson's figure for the number of *known* species alive today (1.4 million) and regard that as an average for the past 550 million years of life, if not the average since the beginning of life 3.5 billion years ago.

Dividing the 3.5 billion years that life has existed on planet Earth by the average species duration of 12.4 million years and then multiplying that value by the number of 1.4 million known extant species, we arrive at a guesstimate of a little less than 400 million for the total number of species aggregated through time. This means that today's known biota of 1.4 million species represents less than four-tenths of one percent of all species that have ever lived. Perhaps 99.6% of all species that have ever lived are now extinct. In the above calculation if we replace the

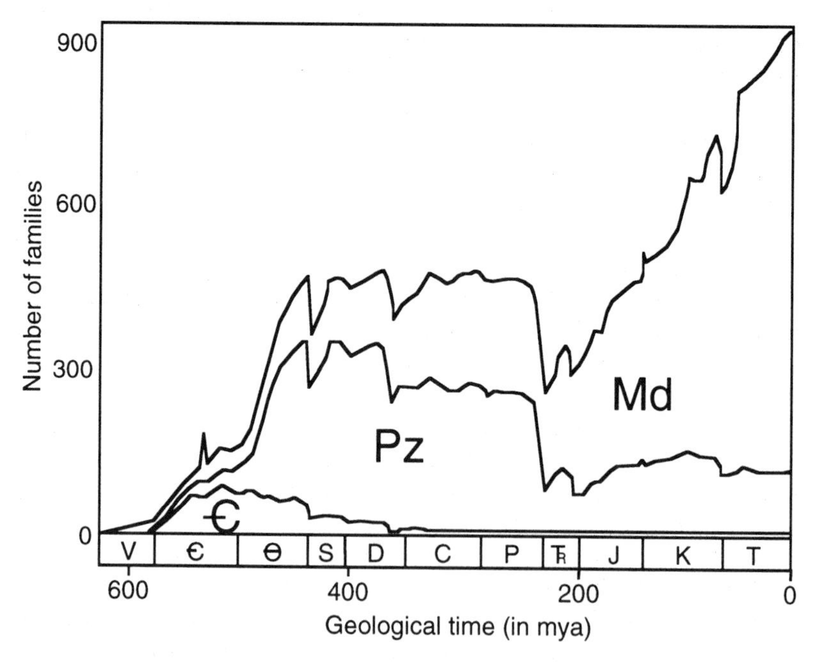

FIGURE 4.1 Animal diversity through Phanerozoic time. The diversity curve shows two major jumps in numbers of animal families since the beginning of the Phanerozoic some 550 mya. The first spurt occurred during the Cambrian/Ordovician; the second, near the beginning of the Jurassic some 200 mya. The contributions of the three major evolutionary faunas to each spurt in diversity are also shown (Є= Cambrian; Pz = Paleozoic; Md = Mesozoic). After Sepkoski and Miller 1985.

time of the origin of life at 3.5 billion years by the estimate of 550 million years for the beginning of the Phanerozoic era, which is the time since organisms with preservable hard parts exploded on the scene, we still find that 98% of all species ever living are extinct.

If nothing else, these rather astonishing figures confront us with the fact that extinction has played a tremendously important role during the history of life. This realization should not, however, provide us any comfort in our current role as the all-time most significant *biological* cause of extinction.

Just as individual humans have a finite existence, so too do species. As shown in figure 4.1, after an initial burst a plateau in the number of species was reached early in the Phanerozoic almost 500 million years ago. That plateau lasted for about 300 million years. Species appeared and disappeared from the fossil record, but the absolute number alive at any one time stayed about the same. As species became extinct, new ones arose—thus balancing losses and gains. A dynamic equilibrium was reached.

Thus in the context of geological time, extinction, like the death of an individual, is the rule—not the exception. The number of people alive today is thought to be greater than the sum of all the people that have ever lived. But this astonishing fact about ourselves should not blind us to the workings of time. The vast majority of species that have ever lived are now extinct. If one accepts the importance of extinction as an evolutionary driving force throughout the paleontologic past, as I do, then being able to correctly recognize extinction in the fossil record is critically important.

SEPARATING TRUE EXTINCTION FROM ARTIFACTS

Recognizing extinction in the fossil record is not as easy a task as it might first seem. Simply identifying the highest (and presumably youngest) record of a particular species in a given region does not assure that the extinction event for this species has been located. As noted earlier, I define extinction as the total loss of a lineage or species and its unique combination of genetic information. This is a fairly standard view of extinction, but it does pose problems for paleontologists. Detecting total loss in the fossil record is not always easy. Looking for biologically meaningful absence is a lot more difficult than looking for presence. In fact, I would go as far as to say that in most instances extinction cannot be detected in the fossil record with great accuracy—despite our valiant attempts. The problem rests with the inadequacies of the fossil record.

In the modern biota we have a fair idea of the distribution of many common plants and animals. If we conclude that all individuals of one of these species dies, we call it extinction. This statement presumes we have detected and accounted for all individuals of the species. The more geographically and ecologically restricted the species is, the more confident we can be of a proclaimed obituary. The wider the ecological tolerance and the greater the geographic range, the less we can trust an assessment of a species' extinction.

For many fossil species the geographic data bank on which we can draw is much more limited than for extant species. The better the geographic coverage, the better we can estimate when a species disappears from the fossil record. To discern extinction in the manner and with the confidence we apply to the modern biota, we would need to consult a fossil record for the entire globe over an interval measurable in tens or hundreds of years. Such does not exist. Instead we assume that in broad terms, when a given morphology (which we take to be a single species) disappears from the fossil record, an extinction has been detected. But is such an assumption justified? Over and over, the fossil record has shown us that the answer is no.

When a species disappears from the fossil record in a particular area, there are four possible explanations. First, the disappearance in one area may be the sign of a true extinction, with the loss of all individuals of that species. Second, the disappearance could be just a local event; individuals of the same species may be known elsewhere from sediments of a younger age. Third, a species may quite simply have evolved into something else. Fourth, the disappearance of a species from that locality may be a vagary of the fossil record caused by poor preservation or inadequate sampling, especially when taxa are rare.

In only the first instance are we dealing with extinction as a total loss. This is what I term *true extinction*. In order to investigate past true extinction, disappearances in the fossil record cannot be taken as evidence of true extinction. We first must tease apart the fossil record to determine if what we see is in fact true extinction, some other biological phenomenon, or whether it is merely an artifact of an incomplete record. Is the apparent disappearance an artifact of local extirpation, of pseudoextinction, or of the vagaries of the fossil record?

Artifact 1: Local Extirpation

One of the very tough questions facing conservation biology today is what should be saved and what should be let go. These are horrible decisions to make, and the resources for determining priorities are very limited indeed. One approach that may help prevent the total loss of individuals of any given species is to preserve the remaining individuals in a protected reserve or, less desirably, in zoos. Genetic variation may be sacrificed along the way—a sacrifice that may bode ill for the long-term viability of a species—but arguably the species is saved, at least for a time.

Although humans today are by far the greatest cause of the local disappearances of species, or what some ecologists call local extirpation, such disappearances have occurred throughout evolutionary history. The causal agents in those prehistoric times were varied, but the single greatest factor seems to have been modification of the habitat of the species. Such modification may become manifest in any number of ways—for example, loss of food sources, changes in rainfall, changes in temperature, loss of nesting sites. In the time of the dinosaurs there were no humans to work their tragic magic of habitat destruction; nevertheless, we see time and time again in the fossil record the local loss of species. I examine this issue starting with a hypothetical case, moving on to two real cases.

Suppose you are in the field, collecting a particular kind of fossil fish. As you move up the rock exposure, you come to a point where the rock changes from shale to sandstone, and you notice that your fossil fish disappears (left column in figure 4.2). Has your fossil fish become extinct—that is, truly extinct?

Thanks to the work of your geochronologic colleague, who has dated a particular volcanic ash in all the sections in the region to approximately 40 mya, you are confident that you can use this ash to find beds of the same age in another location. So you move your investigations a hundred miles (right column in figure 4.2). Here, once again, you find your fish fossil, but this time in shale above the volcanic ash. The loss of the fish at the top of your first rock section is thus not an extinction, but a local disappearance or extirpation caused in this case by a habitat shift.

A reasonable explanation for this simple example would be that as one moved laterally to another geographic expression of the same time, one was seeing the shift and hence loss of marine shales that preserved the fossil fishes. The appearance of sandstone at the top of the first geological section heralded the return of continental sediments, such as deposited by rivers. The marine fishes could no longer live in that area. But they could easily persist beyond the encroaching land.

Events can be more complicated than this simple hypothetical case if the change in habitat and fossils is not also preserved as a change in the kind of rock. Sometimes one needs to investigate the surrounding

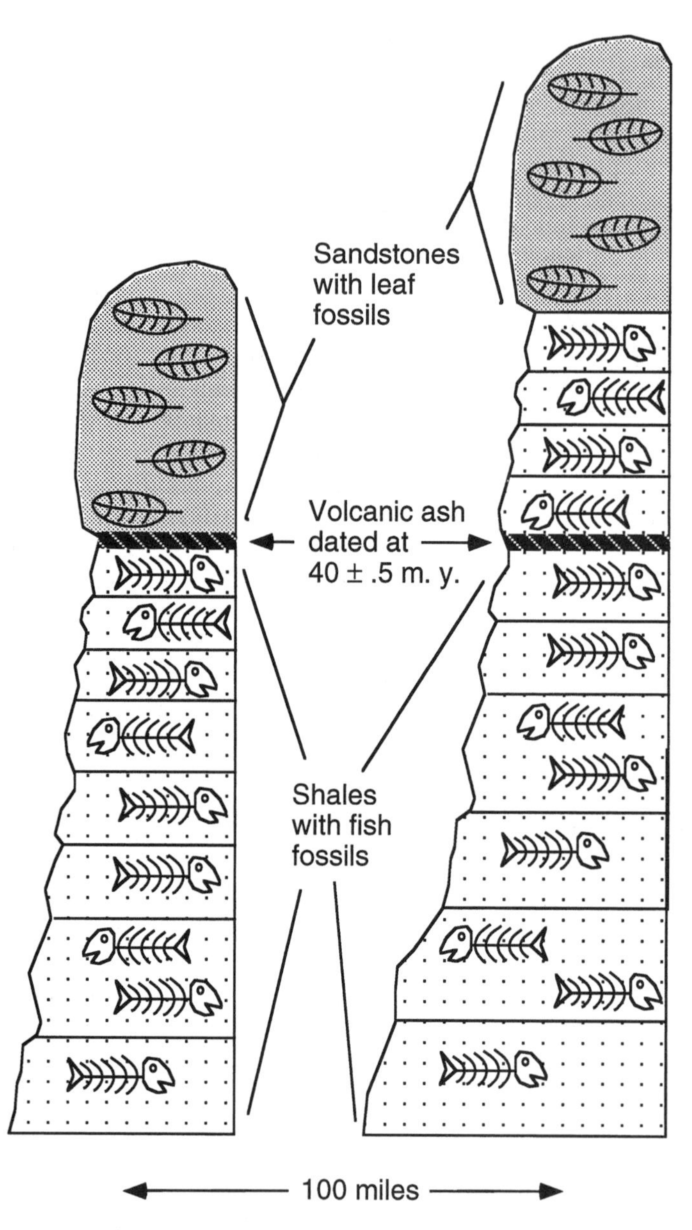

FIGURE 4.2 Local extirpation v. true extinctions. This hypothetical case shows the diachronous disappearance of a fossil fish fauna and the appearance of a fossil leaf flora caused by an environmental shift from a depositional regime that produces shale (light dot pattern) to one that produces sandstone (dense dot pattern). The volcanic ash layer allows the two widely separated outcrops to be correlated in time. We thus can see that the loss of fish in the left section is a local disappearance, or extirpation, caused in this case by a habitat shift. It is a not a true extinction event.

rocks and the habitats that they represent in order to unravel what may have caused the disappearance of a particular species. The disappearance of sharks from the fossil record in eastern Montana at the time of dinosaur extinction is a case in point. And it is a case I am intimately familiar with, as I shall discuss in chapter six.

We know that today some sharks and shark relatives travel upriver a considerable distance from the marine realm. In latest Cretaceous, such a realm lay some fifty to a hundred miles east of the areas I was studying/sampling in eastern Montana. As the Cretaceous closed, the seas retreated far to the south into Texas, severing the marine ties for sharks and their relatives. The sharks vanished from my area of field study. This example is regional in its scope, but the process of extirpation can be continental or even hemisphere-wide in its magnitude and still qualify as a "local" artifact.

Arguably the classic case of hemisphere-wide extirpation is the disappearance of equids (horses) from all of the Western Hemisphere. Equids first appear in the fossil record in western North America some 55 mya. Indeed, this lineage seems to have arisen in North America. Migrations of equids (and other mammals) occurred between the Old and New World at various times during the past fifty-five million years, until their extirpation from the Western Hemisphere just ten thousand years ago (MacFadden 1992, figure 7.7). Horses did not reappear in the New World until reintroduced by Spaniards in the early 1500s. Suggested reasons for extirpation of equids in the New World vary from overhunting by early human immigrants to climatic changes.

This phenomenon of the same species or same higher taxon disappearing at different times around the globe was recognized early by geologists and paleontologists. Disappearance is thus not as sound a marker as first appearance for blocking out chunks of geological time. This rule of thumb is usually followed today. Those who contend that extinctions should be used to mark the K/T boundary should reflect on the vagaries of the equid lineage.

Artifact 2: Pseudoextinction

The appearance of new species by way of evolution (as compared to immigration) is appropriately called *speciation*. If this occurs at the right place and time, and if the species leaves fossilizable hard parts, we have a chance of seeing the record of its genesis and subsequent geographic spread preserved in the fossil record. What happens to the old species whence springs the new? There are only two possibilities—the ancestor disappears or it does not. Both are also within the realm of biological possibility.

If the ancestor does not disappear as its descendant appears, it simply means that some sort of splitting event has occurred. The split could be between populations of approximately equal size. In such a case only one population shows detectable change (Y in figure 4.3A), and it is thus recognized as new while the other remains unchanged (X in figure 4.3A). This is a *bifurcation* (i.e., an equal splitting). Another possibility is that the descendant began as a much smaller subpopulation, with only some of the morphologic and genetic variation of the much larger ancestral species. This is *budding* (figure 4.3B). A further wrinkle is how fast either bifurcation or budding might be. Both a gradual and a punctuated split are possible for bifurcation and budding. Although a fascinating issue (as in Otte and Endler's 1989 book), mode of speciation is not of concern here; rather, what happens to the ancestor is. It is the fate of the ancestor that influences our perception of extinction.

In both bifurcation and budding, which are collectively called *cladogenesis* because a new clade or clades are formed, the ancestral species need not disappear from the fossil record. There are, however, two possible forms of speciation in which the ancestral species does disappear. If in the process of bifurcation both resulting species change, then the ancestral species disappears. We could mistakenly identify the disappearance of species X in figure 4.3E or F as an extinction, when it was really a pseudoextinction. A *pseudoextinction* in this case means that one lineage splits into several lineages. Species X is gone, but species Y and Z take its place—a net gain in biodiversity attributable to the pseudoextinction.

A second possible pseudoextinction occurs when the ancestral species evolves into a descendant species without bifurcation (figure 4.3G and H). This is termed *anagenesis*. The process driving this kind of speciation, as with all others, is uncertain. Whatever the process, however, the disappearance of the ancestral species is a pseudoextinction because there is genomic continuity between ancestor and descendant, unlike true extinction that would eliminate the entire genomic line.

Overall, in order to separate true from pseudoextinction in the fossil record, we need a reasonable hypothesis of how the species in question are related. Without this we really cannot go beyond simply saying that a particular named species, genus, or other higher grouping disappears at thus and such a time and place in the fossil record. I recently addressed this issue, using early Tertiary mammals in North

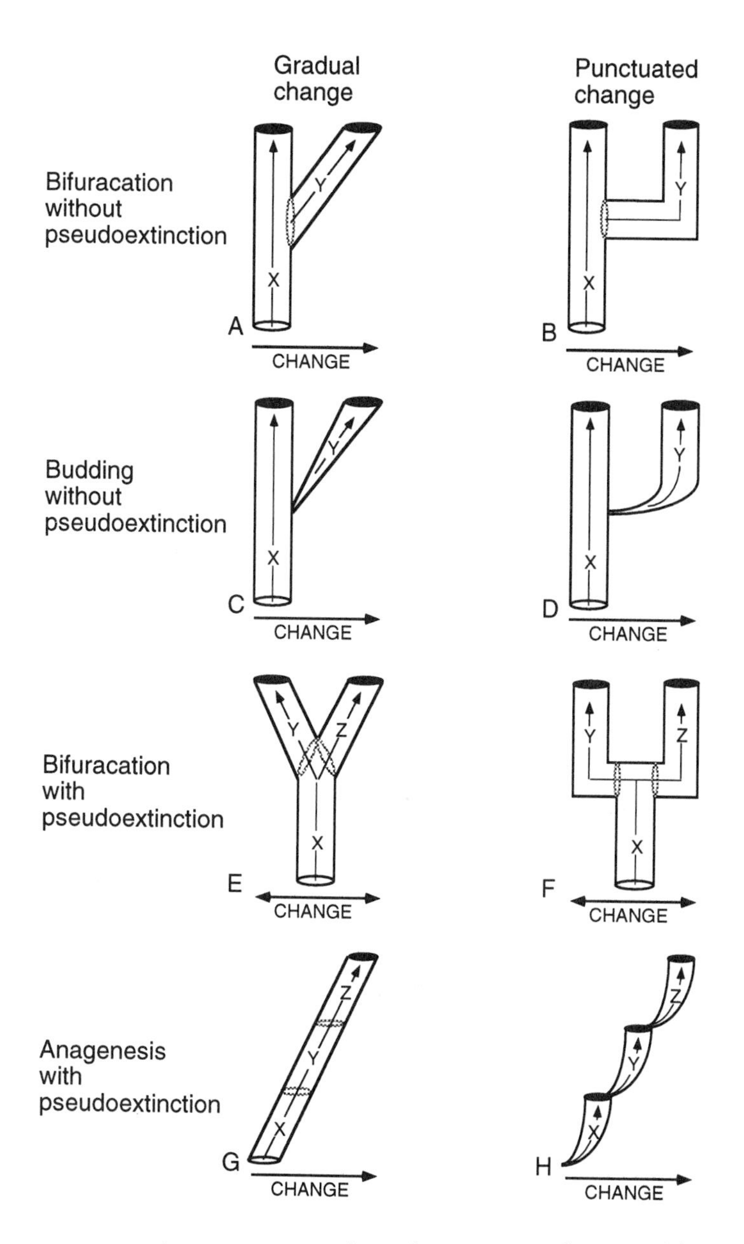

FIGURE 4.3 Evolutionary causes of pseudoextinction. There are eight possible ways in which species X might be affected over time, short of outright extinction. In (A) through (D) species X remains intact, while giving rise to species Y. In (E) through (H) species X changes into species Y and Z; in these cases the disappearance of species X is not a true extinction—not an outright loss—but a pseudoextinction. Relative population size is approximated by size of the cylinder representing the clade, and the shift to the right or left represents change—whether in morphology, behavior, or genes. If an ancestral species (X) bifurcates (A and B) or buds (C and D), but does not itself change, it may be found in the fossil record after speciation. If, however, an ancestral species bifurcates into two new species (E and F) or changes without bifurcation into a new form (G and H), the ancestral species disappears from the fossil record because of pseudoextinction. After Archibald 1993a.

America (Archibald 1993a). Although the biochronological placement of the mammalian fossils—that is, their relative placement in geological time—is moderately well known (Archibald et al. 1987), only three of the major lineages had been studied phylogenetically at the species level. This is not the best quality fossil record to use in asking questions of pseudoextinction versus true extinction. A later Tertiary sample of mammals or, better yet, marine invertebrates might offer a more nearly complete fossil record. Nevertheless, even with these caveats, for the three major North American mammal lineages of the early Tertiary (including 31 species) I was able to examine, a surprising 25% of the disappearances were pseudoextinctions of the kinds shown in figure 4.3.

If pseudoextinctions are not looked for and distinguished from true extinctions, the level of mammalian extinction calculated at the K/T boundary could well be inflated by almost 50% (chapter 7). Without further studies, we do not know the degree to which pseudoextinction may be skewing our perception of true extinction in the geological past. I do feel, however, that it is a long-neglected issue.

Artifact 3: Vagaries of the Fossil Record

Far more of the record of Earth's evolutionary past has forever been lost to the forces of decomposition, alteration, and erosion than we shall ever come to know. Even if we could miraculously see everything now hidden in the rocks, the vagaries of preservation would deny us the full picture. The spotty nature of the geological record is not unique to natural history. All histories bear this burden. There is, however, no cause for despair. The information that has been preserved in the rock and retrieved by human effort is truly a wondrous précis of past life. The difficulty comes when we must determine whether our record is accurately portraying the biological past.

Four patterns owing to the vagaries of the fossil record are of particular concern when dealing with biotic events at the K/T boundary. Two have names—the Lazarus Effect and the Signor-Lipps Effect. To these I add names for problems long recognized but lacking proper names—the Zombie Effect and the Rarity Effect.

The *Lazarus Effect*, coined by David Jablonski, refers to a taxon that has a gap in its record, hence the allusion to the raising or reappearance of the presumed dead (Flessa and Jablonski 1983). There may be several reasons for such a gap. Consider the hypothetical example shown in figure 4.4, the disappearance of a species of fish. The disappearance of species A coincides with a change in environment, repre-

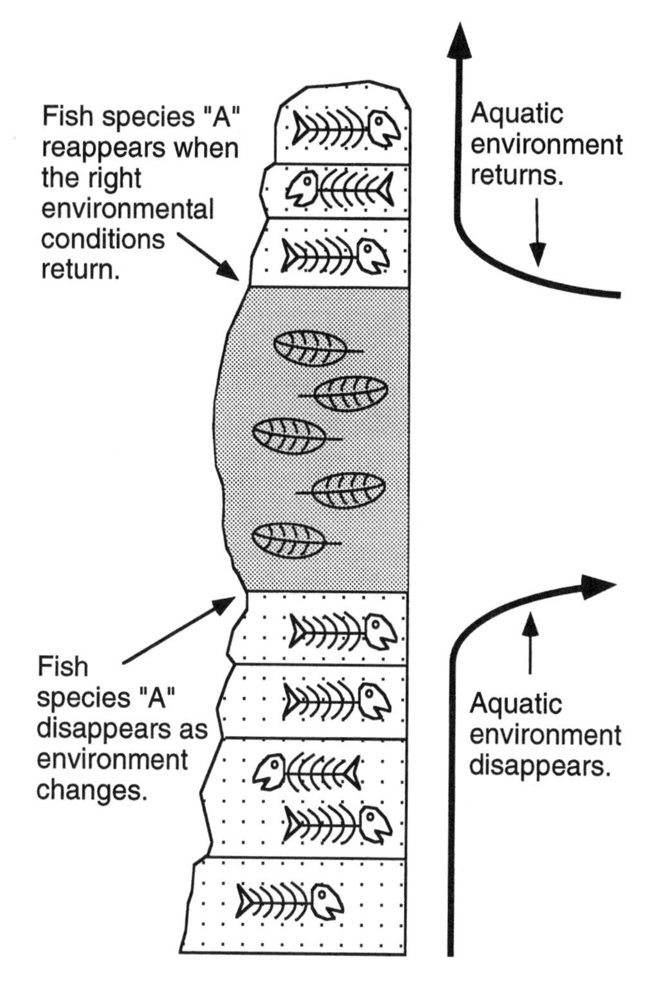

FIGURE 4.4 The Lazarus Effect. The Lazarus Effect (named in Flessa and Jablonski 1983) occurs when a taxon has a gap in its record, hence the allusion to rising from the dead. There may be several reasons for such a gap. In the hypothetical rock exposure or formation shown here, a particular fish species disappears because there is a change in environment, but it returns when the original environment is reestablished.

sented in a new kind of rock. When the original environment returns, the fish species, "A," also returns. If no important morphological changes have occurred during the time lapse, then continuity is clear.

If taxa burdened by the Lazarus Effect are not pulled out and given special attention, the number of extinctions could be overestimated. For example, in a study of Late Cretaceous through mid-Paleocene vertebrate faunas, Laurie Bryant and I (Archibald and Bryant 1990) found that as many as ten genera known in the latest Cretaceous Lancian faunas seem to disappear at the K/T boundary—only to reappear in the middle Paleocene, Torrejonian faunas (57 versus 67 genera in table 4.1). If each of these genera is monophyletic (that is, if all species grouped as such have a single common ancestor that is shared by no member of any other genus), then the reappearance of the same or new species belonging to that genus would be an example of the Lazarus Effect. But the Lazarus Effect vanishes if the reappearing members of the genus have been wrongly classified into the genus of the earlier appearing species. We must thus be cautious that higher taxa evidencing disappearing and reappearing acts are not catch-all wastebasket genera that group unrelated species.

The *Signor-Lipps Effect*, named for the two invertebrate paleontologists who recognized it (Signor and Lipps 1982), is more subtle than the Lazarus Effect and thus potentially more difficult to detect. Suppose that as we approach a boundary at which an extinction event occurred, we note that the number of specimens and number of species diminishes. Are these diminutions real or artifacts? A possible sampling bias might be that the number of sampling sites diminishes as the boundary is approached (Flessa and Jablonski 1983). Signor and Lipps (1982:291) argued, "If the distribution of last occurrences is random with respect to actual biotic extinction, then apparent extinctions will begin well before a mass extinction and will gradually increase in frequency until the mass extinction event, thus giving the appearance of a gradual extinction." There is no basis to assume that there should be an increase in frequency of apparent extinctions as the mass extinction is approached, but because of the vagaries of preservation I think Signor and Lipps correctly conclude that extinction events may appear more gradual than they really are.

For example, in figure 4.5A note that ten species are variously preserved throughout the entire geological section. Such a spotty distribution is quite possible because of ecological changes. For simplicity I have chosen a regular pattern of occurrence and disappearance for each of the ten species. Notice that if one had the time to sample and thus to view the geological column in full, there would be no question

Table 4.1 The Lazarus Effect Across the K/T Boundary

Without accounting for the Lazarus Effect

	Judithian	Lancian	Puercan	Torrejonian
Judithian	101	—> 55 (54%)	—> 23 (23%)	—> 18 (18%)
Lancian		97	—> 38 (39%)	—> 24 (24%)
Puercan			57	—> 29 (51%)

Accounting for the Lazarus Effect

	Judithian	Lancian	Puercan	Torrejonian
Judithian	101	—> 56 (55%)	—> 29 (29%)	– – –
Lancian		99	—> 48 (48%)	—> 30 (30%)
Puercan			67	—> 39 (58%)

Source: After Archibald and Bryant 1990.

Notes: Data are for North American vertebrates from the latest Cretaceous (Judithian and Lancian) through early (Puercan) and mid (Torrejonian) Paleocene. The numbers in larger font are the total numbers of genera present during each time interval. Numbers in smaller font are the number of genera and the percentage of genera surviving from the previous interval. For example, without accounting for the Lazarus Effect, 55 of 101 genera (54%) present in the Lancian were survivors from the previous Judithian (the rest were new or immigrant genera); 23 of 101 genera (23%) present in the Puercan were survivors that had held on since the Judithian, whereas 38 of 97 genera (39%) present in the Puercan had carried over from the immediately preceding Lancian.

that all ten species were doing business as usual right up to the extinction boundary. Yet because of the happenstance of sampling choice, a gradual extinction is what one perceives. The four equally spaced sampling intervals suggest that there is a gradual reduction from ten, to six, to four, and finally to three species at the extinction event, even though a full view of the rock record shows how our sampling choice would have misled us. Such an even distribution of fossils and such an unfortunate sampling choice are possible, but as the numbers of localities are increased, the vagaries of sampling would likely diminish.

The distributions in figures 4.5B and C are more realistic for vertebrates. The vast majority of fossil vertebrates are found at localities or at specific horizons. The intervening rock record is often a virtual blank. Here it would do no good to randomly determine intervals to sample; one simply takes what one can get and then tries to ascertain trends. The distribution of fossils is not random, although there may not be any easily discernible regularity in the stratigraphic placement of localities. If we see no change in the number of species proceeding up section (figure 4.5B), and especially if the fourth and final expression of fossils occurs right below the boundary, then a catastrophic mass extinction is the best explanation.

In C, the number of species declines at localities going up section. A gradual mass extinction is thus plausible. As the number of geological sections preserving such localities increases, and if the pattern still holds, a gradualistic interpretation becomes stronger and stronger. The pattern seen in C is what we see over and over on a regional scale for the "ten-foot gap" of dinosaurs below the K/T boundary in the Western Interior (discussed in chapter 3). The gap is real, but it probably owes to something other than extinction—possibly lack of preservation or dissolution of bone.

The third artifact caused by vagaries of the fossil record I term the *Zombie Effect*. The implication is that sometimes fossils from a given locality may actually be reworked by either biological or geological processes from older rocks. Thus fossils that at first seem to be the remains of organisms that lived at the time of the formation of the fossil locality are actually the exhumed remains of organisms that lived earlier. They lurk in later sediments like the living dead. In fossil localities formed in quiet water, the major source of the Zombie Effect is bioturbation caused by the reworking of older sediments by living organisms—usually small to microscopic invertebrates that burrow into or forage within the sediments. In fossil localities formed by more active transport, such as stream deposits, the reworking process is usually mechanical.

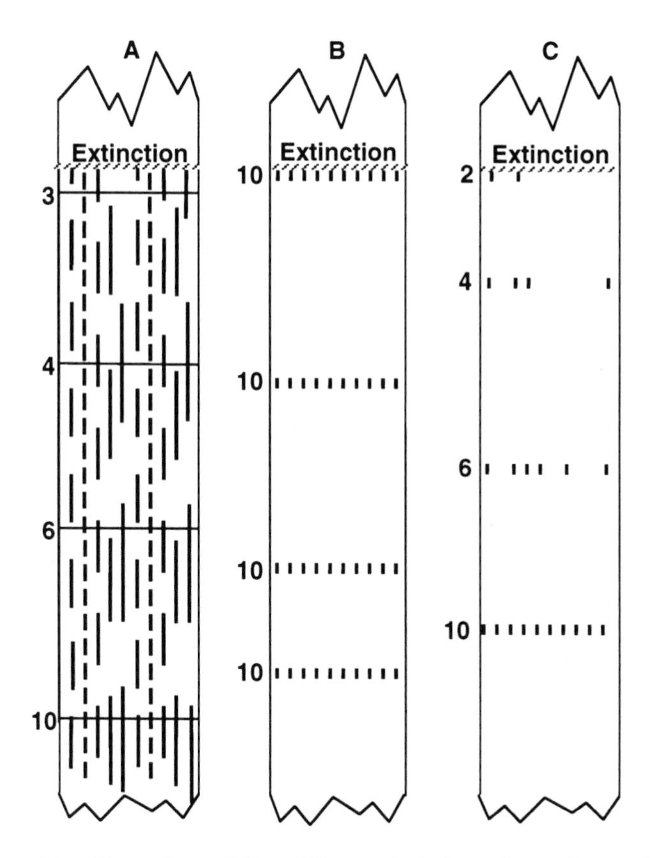

FIGURE 4.5 The Signor-Lipps Effect. The statistical tendency for an abrupt extinction in real life to appear gradual in the fossil record was noted by Signor and Lipps (1982). The geological section in (A) shows ten species, each with a distinct and regular pattern of preservation. (Absolute regularity would be highly unnatural, but the regularity here does show how sampling can lead to confusion.) The results drawn by taking samples at just the four intervals shown would suggest a gradual reduction from ten to six to four to three and finally to total species extinction. In reality all species probably made it to the extinction event. For geological sections (B) and (C) we have a more realistic pattern for large samples of fossil vertebrates, with most found at localities or specific horizons. Fossil distribution is not random, but the pattern may not be discernible. In (B) our sampling detects no change in species numbers right on up to the (presumed) catastrophic extinction; confidence in sudden extinction is thus high. Section (C), however, suggests gradual extinction if sampled over many geographic regions; this Signor-Lipps Effect can be all but eliminated.

Imagine a shallow stream. As it flows, it cuts away at sediment in the banks along the outside of curves. It then redeposits those excavated sediments downstream in the inside of the curves, forming sandbars. The sediment that is cut away may well contain the remains of plants and animals that lived on the floodplain tens, hundreds, or even thousands of years earlier. As the older remains are carried or pushed along in the stream, these zombies may become mixed with recently dead plants and animals. The mixed remains may then be deposited on one of the sandbars, quickly buried by a flood, and preserved. If the remains are later excavated by curious humans we have a Zombie Effect that might influence our perception of what the flora and fauna was like at the time of deposition of the sandbar.

I argue that the Zombie Effect is far more common than we paleontologists would like to believe. Basically, we should be suspect of any fossil bone that is not articulated with a good bit of the rest of the skeleton. Fortunately, however, in most cases the reworking occurs in a short enough time span that the reworked remains are from plant and animal species that are still extant at the time of the exhumation of the older plants and animals. The real dilemma arises when some major event, such as at the K/T boundary, occurs during this process of exhumation and reburial. Thus, when we hear about dinosaurs existing past the K/T boundary, we should wonder if these are reworked specimens—Zombie taxa.

There is another interesting potential manifestation of the Zombie Effect that seems to have gone unrecognized. In the case of the K/T boundary we usually worry that fossil specimens that appear above the boundary may have been reworked from Cretaceous rocks, thus giving the false impression that these species survived into the Tertiary. But we should bear in mind that it should be just as common for older, already extinct species to be reworked into higher sediments that are nevertheless still below the boundary. This would give the impression that species became extinct much closer to the K/T boundary than they really did. In this case then, the Zombie Effect is the flip side of the Signor-Lipps Effect; Zombie taxa lead to the overestimation of extinction in a given interval or at given level, while the Signor-Lipps Effect would have us underestimate extinction.

Figure 4.6 shows a hypothetical case to illustrate the Zombie Effect. On the upper right is a horizon where there appears to be the mass extinction of a large number of species, here for simplicity represented by only five shapes—square, circle, triangle, ellipse, and rectangle. When the sedimentological setting is investigated, it becomes clear that a succession of five channels have cut down and cut laterally into older sediments (as indicated by the overlap of the channel deposits) during the process of filling our hypothetical basin. Starting at the bottom channel, we find fossils of the rectangle and ellipse. The next higher channel deposit contains four shapes. By the third channel deposit we have all five shapes. This pattern continues through the top channel. But because of the reworking aspect of these channels we do not know that from the third through fifth channel some of the shapes had become extinct (unfilled shapes) and that they were reworked as Zombie taxa from older channel deposits. In this case our mass extinction is completely illusory. As these Zombie taxa are reworked several times, their numbers would begin to dwindle as specimens are destroyed and not reburied. This is a possible reason why we see the dwindling numbers of dinosaur teeth and other latest Cretaceous vertebrates as we go to younger and younger localities within the Bug Creek sequence in eastern Montana.

The final artifact, long recognized but which I like to call the *Rarity Effect*, can be viewed not only as a vagary of the fossil record but also as the rarity of some species. For example, when one of my cats catches a hapless bird in my yard (even though the cat is belled to warn birds), chances are that the victim will be a house finch, dove, or mocking bird rather than a California towhee or Cedar waxwing. Similarly, the chances of finding a fossil of a particular species reflects the rarity of that species when extant. If we have what appear to be very rare species in the Cretaceous Hell Creek Formation, but find none of them in the overlying Paleocene Tullock Formation, does it mean the species became extinct or is simply so rare that we have not yet sampled it? The answer, unsatisfying as it may be, is both are possible. I delay discussing this very tough issue until chapter 6, where it becomes an important consideration for understanding the fate of the vertebrates at the K/T boundary.

PUNCTUATIONS IN THE HISTORY OF LIFE

It should be clear from the preceding discussion that discerning the signal of true extinction from various kinds of noise caused by vagaries of the fossil record is not an easy task. If nothing else, logic tells us that extinction has, does, and will continue to occur; for paleobiologists it is the rule and not the exception. Most of us also agree

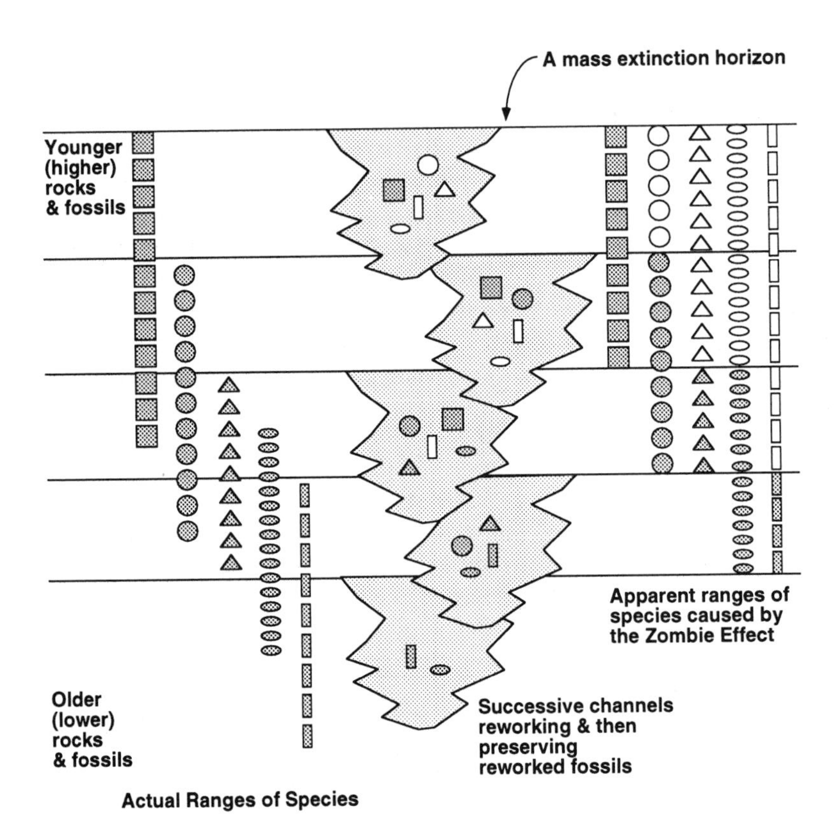

FIGURE 4.6 The Zombie Effect. In this hypothetical case we know the actual distribution of species in time (the filled shapes on the left). Channels preserve the remains of organisms living at the time of channeling (filled shapes in channels). Channeling may also rework and then redeposit fossils of extinct species into younger sediments (outlines of shapes in channels). Such are the living dead, or zombie, species. The Zombie Effect occurs when zombie species mislead us into concluding that the extinction happened later than it actually did, such as at the top of columns to the right.

that the pattern of diversification of life through the last 550 million years follows something like the curve in figure 4.1. The next step is to ask whether we can discern extinction patterns. If so, what are these patterns?

Discerning Patterns—Mass v. Background Extinction

The two most important patterns generally read from the fossil record are: (1) A background of normal levels of extinction has been present since the beginning of the Phanerozoic some 550 mya, if not starting with the origin of life some 3.5 billion years ago. (2) The background, or normal, extinctions have been punctuated at various times by much higher levels of extinction; these are the so-called mass extinctions.

A third major pattern of extinction merits brief comment because it is important in any general discussion of extinction patterns. It is, however, only tangential to the specific subject of K/T extinctions. This pattern, claimed by David Raup and Jack Sepkoski (1986), is that there is a periodicity of twenty-six million years for mass extinctions. The claim of periodicity has not found wide support among paleobiologists, for lack of a demonstrable cause and issues concerning the methods of tabulation. For those interested in further reading, Doug Erwin presents a moderately critical but still sympathetic review of the periodicity controversy in his 1993 book on the terminal Permian extinctions; Andrew Smith explains some of the problems with this approach in his 1994 book on the importance of systematics in the fossil record.

Returning to the first two patterns, that of normal background extinction punctuated by mass extinctions, one of the most widely cited papers is also by David Raup and Jack Sepkoski (1982). From this paper comes the most copied illustration showing the difference between background and mass extinction. I repeat it here in figure 4.7, in a slightly simplified version. The graph shows total extinction (extinctions per million years) through time for families of marine invertebrates and vertebrates. It does not pertain to the terrestrial realm. The dots in the lower portion of the graph are the levels of background extinctions through the past 550 million years, while the peaks (Late Ordovician, Late Permian, Late Triassic, and Late Cretaceous) are mass extinctions that Raup and Sepkoski found to be statistically different from background. A fifth mass extinction in the Late Devonian is possible, but it is not confirmed in a statistically significant way. Together these mass extinctions constitute what the

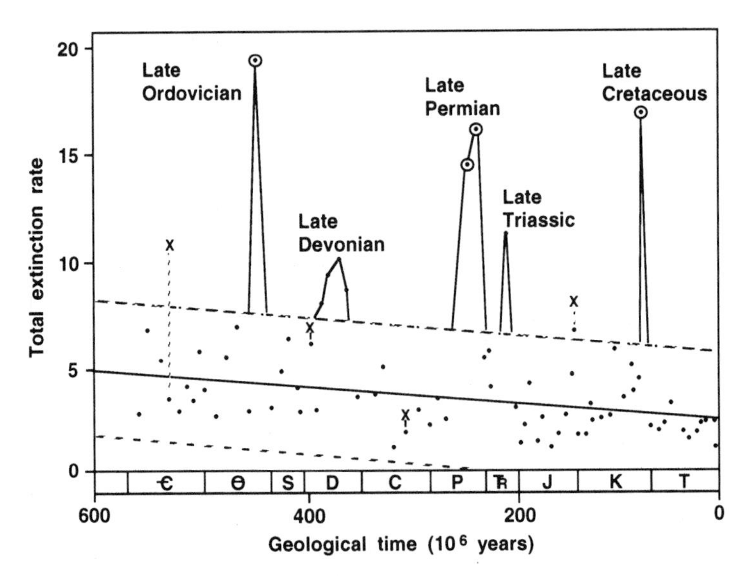

FIGURE 4.7 The so-called Big Five extinctions. This simplified version of perhaps the best known figure showing alleged mass extinctions expresses the number of family-level extinctions per million years. Calculations were drawn from data for marine invertebrates and marine vertebrates throughout the Phanerozoic. Four of the five (Late Ordovician, Late Permian, Late Triassic, and Late Cretaceous) are mass extinctions statistically different from background. The Late Devonian mass extinction is apparent here, but it is not statistically significant. After Raup and Sepkoski 1982.

champion of impact extinction, David Raup (1991), has called the "Big Five" mass extinctions.

Reality or Artifact?

The distinctions between mass and background extinctions seem reasonable and straightforward, but these results have not gone unchallenged. The most critical issue is that the methodology has a built-in taxonomic artifact that may greatly exaggerate the level of extinction. This is because some taxa that constitute the database may not be monophyletic. A monophyletic taxon is thus one that includes all known descendants and a common ancestor not shared with other such groups. In a very real sense it is the only nonarbitrary kind of taxonomic group. In the tabulations of Raup and Sepkoski (1982, 1986), distinguishing monophyletic from nonmonophyletic groups would be a tricky effort. Nevertheless, the issue cannot be ignored.

In a series of papers, Colin Patterson and Andrew Smith used a subset (fishes and echinoderms) of the Raup/Sepkoski data set to show the importance of caution in using compilations of taxonomic names when assessing extinction. In a 1994 book, which references the Patterson and Smith papers, Andrew Smith reviews this issue and also raises the further concern of using higher taxa (such as the families used by Raup and Sepkoski) as proxies for species. In their analyses Smith and Patterson found that about 20% of disappearances among fishes and echinoderms during mass extinctions were taxonomic noise of the kinds just mentioned. These results parallel the 25% level of pseudoextinction for early Tertiary mammals, which I mentioned earlier in this chapter.

If the mass extinction peaks of Raup and Sepkoski shown in figure 4.7 are 20% artifact, does this mean these were not mass extinctions? I do not know if such a drop would render the four statistically significant peaks nonsignificant, but some sort of peak would still exist at least for the four. Thus the noticeably higher, even if not significantly higher levels of disappearance, tabulated by such authors and Raup and Sepkoski do appear to be very real.

Mass Mortality v. Mass Extinction

We have all read in the papers or have seen television reports of the mysterious beachings of whales or mass deaths of birds and fishes. Mass mortality is the death of tens to possibly billions of organisms, yet in no cases that I know have these mass mortalities led to mass extinction or even regular background extinction. What then, if anything, do mass extinctions and mass mortality have to do with one another. I would argue "little, if anything."

I draw attention to this issue because in the question periods following talks I have been asked about mass mortality and mass extinction. The only circumstance in which one might legitimately make a connection between mass mortality and mass extinction is if a very widespread catastrophic event kills virtually all members of a species. Some, but not all corollaries of an asteroid or comet impact, such as that which is argued to have struck Earth sixty-five million years ago, would have caused mass mortality as well as mass extinction. But, as I will discuss in chapter 7, there is little evidence supporting the thesis that corollaries of impact caused mass mortality.

Even with all the controversies surrounding events at the K/T boundary, no one that I know or read doubts the importance of extinction in the evolutionary process. For me, the controversies (and myths) arise when we try to explain these *biological* events without very careful attention to what the fossils have to say. In the remainder of this book I pay attention to this fossil record—especially the vertebrates—and how the record provides tests for our hypotheses of biotic turnover.

V

Who's Who of the Late Cretaceous

Despite what we read in the papers and see on the silver screen, dinosaurs were not the sole backboned creatures populating the Mesozoic world. By some measures, in fact, they were not even the most important. When dinosaurs are portrayed both in and out of science as the dominant creatures of their day, we are inevitably gauging in human, especially western, frames of reference. We tend to equate dominance with ideas of size, power, and success. Even in the business world, these modifiers can sometimes be difficult to define; when we turn to biology the definitions are even more tenuous.

MEASURING SUCCESS

One way to avoid pitfalls of overgeneralizing about biological dominance or success is to specify how these factors are measured. Some relatively straightforward examples are: sizes of the organisms, the numbers of species in a group, the numbers of individuals in a species or a local population, and how diverse morphologically and thus ecologically a group is. How big, how diverse, how numerous, how morphologically varied were the dinosaurs of the Late Cretaceous, compared with other vertebrates in their biological community? Although I limit my comparison to the latest Cretaceous faunas of eastern Montana, the conclusions drawn from this limited geographic range probably apply reasonably well to other regions and other continents.

As to *body size*, dinosaurs win hands down, at least if we limit the scope of study to adult dinosaurs. Except for the largest turtles, crocodilians, and probably one lizard, all species of adult dinosaur in the latest Cretaceous of eastern Montana are larger than any other vertebrate species of the time.

Judged by *numbers of species*, however, dinosaurs are certainly not dominant. Rather, mammals come out ahead by this measurement. The fossil record can tell us little about the birds of this time; their lighter bones worked a bias against preservation. Eastern Montana has produced just a handful of bird specimens of latest Cretaceous age, assigned to about five species. Five is simply too few to understand what role birds played in the Late Cretaceous, but if we (properly) add these five to the roster of saurischian dinosaurs, then dinosaurs do rival mammals in numbers of species. My suspicion is that in this measure, dinosaurs would be even higher if only the record of Mesozoic birds were not so spotty. Consider the prominence of birds today: birds are the most species-rich group of land vertebrates, with some 9,000 species compared to 4,200 species of mammals. Even today's squamate reptiles (lizards and snakes) are more numerous than mammals, with some 6,000 species. By this count, the Age of Mammals is a misnomer. This is surely the Age of Birds or even the Age of Squamates!

What about total *population size*? Trying to estimate numbers of individuals within a species or higher taxonomic group is difficult even for very large living vertebrates. Ecologists are becoming better at counting numbers of elephants or wildebeest; estimates, however, can vary. It thus comes as no surprise that estimating numbers of individuals of dinosaurs in a species or in a given area is at best a guess. Reports of hundreds of skeletons, such as the ceratopsian *Centrosaurus* in southern Alberta (Currie and Dodson 1984), or even thousands of skeletons—notably, the hadrosaur *Maiasaura* in western Montana (Horner and Gorman 1988)—present a strong case for the herding of large numbers of some dinosaurs. For the vast majority of dinosaurs, however, next to nothing is known of their social structure, including the size of their groups. This tempts us to extrapolate the tidbits of knowledge about population size and behavior to closely related species. But extrapolation can be dangerous. An exercise using the even-toed ungulates alive in Africa today can show why. Species within the same family Bovidae (antelope, cattle, goats, sheep, etc.) may have quite different social structures, and thus markedly different numbers of individuals within a typical group.

The fourth and final measure of dominance, *morphological and ecological diversity*, is probably most easily examined by comparing distinct groups of vertebrates, rather than attempting to establish some numerical measure. At all times during their existence, dinosaurs showed some degree of morphological and ecological diversity. This is especially true during the Late Cretaceous, the time of greatest species diversity. Dinosaurs are more ecologically diverse than most people envision, even if their modern representatives—the birds—are excluded.

Dinosaurs (with or without birds) fall behind *living* mammals in morphological and ecological diversity. This is not an artifact of unbalanced data sets. The difference is real. As vertebrates go, mammals today (and probably as far back as the Eocene epoch) are extraordinarily diverse. They are found in the hottest and coldest places. They thrive on land, in the sea, and in the air. They range in size from a tiny shrew to the largest vertebrate of all time—the blue whale.

Two key factors combined to allow this mammalian diversity: viviparous reproduction and endothermy. *Viviparity* (or live birth) refers to birth after a long period of gestation within the female. One group of mammals, the placentals (or eutherians)—the group to which we belong—has taken this to the extreme by prolonging gestation for up to almost two years in the case of elephants. Extended gestation provides an extremely stable environment in which the embryo can develop and thus frees the mother to enter quite inhospitable environments.

Endothermy is the production of internal heat—body heat. Eighty percent of the caloric value that an average mammal obtains in its diet is used to sustain its proper metabolic level. The remainder is available for growth, motility, reproduction. This is a very expensive way to go, but it has one very great advantage. High metabolism in part decouples the organism from its environment, allowing it to function in colder or even sometimes hotter settings than would otherwise be possible.

No other vertebrates evolved both viviparity and endothermy in combination. Birds are endothermic, but all are egg layers. In fact, all extant Archosauria—birds and crocodilians—are egg layers. This suggests (but does not prove) that dinosaurs, as archosaurs, were also egg layers. But there is some direct fossil evidence of egg laying in a few dinosaurs.

Relatively little evidence connects most species of dinosaur with any specific kind of eggs. The connections that are known show oviparity occurring within both of the two broad groups of dinosaurs.

Among the saurischians, or reptile-hipped dinosaurs, we can be moderately confident that the ponderous sauropods and the carnivorous theropods both had representatives that were egg layers. In deposits of Late Cretaceous age in southern France, large eggs are associated with, but not directly linked to, the remains of the sauropod *Hypselosaurus*. *Troodon*, a small theropod from the Late Cretaceous of North America, is similarly associated with, but not positively linked to, eggs found in the area worked by Jack Horner in western Montana. Among the ornithischians, or bird-hipped dinosaurs, both ornithopods (duck-bills) and ceratopsians (horned dinosaurs) are associated with eggs. The ornithopod *Maiasaura*, studied by Jack Horner, is directly linked with eggs, and a complete growth series of embryos, juveniles, subadults, and adults is also known. The famous eggs from the Flaming Cliffs of Mongolia that had long been matched with the early ceratopsian *Protoceratops* have recently been reinterpreted as belonging to an oviraptorid. An embryo of an oviraptorid was found in a partial egg (Norell et al. 1994). Oviraptorids are beaked, toothless, bird-like saurischians best known from the Gobi Desert of Mongolia.

Endothermy in dinosaurs remains a hotly contested issue. A variety of evidence based on growth rates, bone histology, comparative temperatures throughout the body, comparative anatomy, and trackways has been used to argue that most dinosaurs were endothermic—that is, they were capable of producing their body heat through metabolic means. John Ruben, a physiologist working on both extant and extinct creatures, published a 1995 review article on the evidence for endothermy in dinosaurs. He points out that when we compare extant endotherms (birds and mammals) with extant ectotherms (such as crocodilians and lizards), we find that supposed indicators of endothermy in dinosaurs can also be found in some extant ectotherms. These same indicators are sometimes lacking in extant endotherms. Nevertheless, because most of these indicators of endothermy, whether equivocal or not, appear together in some taxa of dinosaurs, I find it reasonable to argue that at the very least these taxa had levels of activity reaching those of mammals and birds. Even if all dinosaurs lacked the kind or degree of endothermy that birds and mammals evince, there is no denying that simply by virtue of great size many adult dinosaurs must have had a relatively high, relatively constant body temperature. The reason for this higher, more stable temperature owes to a simple change that occurs as objects of a given shape change size.

Picture a cube that has edges an inch long. Each face of the cube is thus one square inch, for a total of six square inches. The volume of this cube is one cubic inch. The ratio of surface area to volume is thus six square inches to one cubic inch, 6:1. Now imagine a cube that is doubled in its linear dimensions, so that each edge is two inches long. Each side has a surface area of four square inches for a total of twenty-four square inches; the volume is eight cubic inches. The ratio of surface area to volume for this larger cube is 24:8, or 3:1. In going from the small cube to the large cube, we have reduced the surface area to volume ratio from 6:1 to 3:1. The large cube will thus absorb or radiate heat much more slowly than will the small cube.

This relationship also applies to animals, and thus a large dinosaur would have lost or gained heat much more slowly than would a small lizard skulking in the underbrush. This greater thermal inertia has very appropriately been named *gigantothermy*. If anything, the behemoths among the extinct Dinosauria would have had difficulty in cooling down, not in staying warm. This is simply the mechanical result of an inability to dissipate heat rapidly because of great mass.

The recent finds of dinosaurs at high latitudes in Alaska (Clemens and Nelms 1993) and in Australia (Vickers-Rich and Rich 1993), which abutted the present position of Antarctica in the late Mesozoic, shows that dinosaurs could withstand cooler, if not cold temperatures. But because of prolonged periods of darkness, they probably migrated seasonally toward slightly lower latitudes in search of food, just as some birds and mammals do today.

One other factor has probably contributed to the greater ecological diversity of Cenozoic mammals. This is what Stephen Jay Gould has called "historical contingency" (1989). Quite simply, some of the most fundamental aspects of the pathways that any particular group of organisms takes during its evolution are largely a measure of where they have been evolutionarily. For example, the ancestor of all modern mammals was probably a small, generalized quadruped with partially grasping digits. The ancestor of all dinosaurs (including birds) was probably an animal that had already begun to reduce digits and emphasize bipedality. At least three and possibly more times dinosaurs returned to quadrupedality rather than sticking with the bipedal tendency, but much of this trend was probably driven by the demands of increasing size. Carrying oneself on all fours is a good way to handle great bulk. Thus, compared to mammals at least, subsequent dinosaurian evolution exhibits more channeling of what was possible in locomotion patterns.

Body size, numbers of species, population size, and morphological and ecological diversity are by no means the only measures of evolutionary success or dominance. But they are those that come to mind when we look at Mesozoic dinosaurs, especially compared to Cenozoic mammals. The rest of this chapter will survey the vertebrate fauna of eastern Montana's Hell Creek, group by group. By comparing the dinosaurs with their own comrades in time—not with today's vertebrate contingent—the dinosaurian story becomes even more apparent.

A SURVEY OF THE NONARCHOSAURS

An essential place to begin a discussion of the vertebrate contemporaries of the Late Cretaceous dinosaurs is with some systematics. How do the major groups relate to one another evolutionarily? Let's take another look at the first cladogram that appeared in chapter 2 (figure 5.1). It is a very simplified version of vertebrate relationship, leaving out many major groups, but it does show how the important groups of Late Cretaceous vertebrates relate one to another.

Recall that this (and any other cladogram) is constructed so that members of each clade, and at each higher level of grouping, all share a more common recent ancestor with one another than any does with a group of animals outside the clade, extant or extinct. This means, for

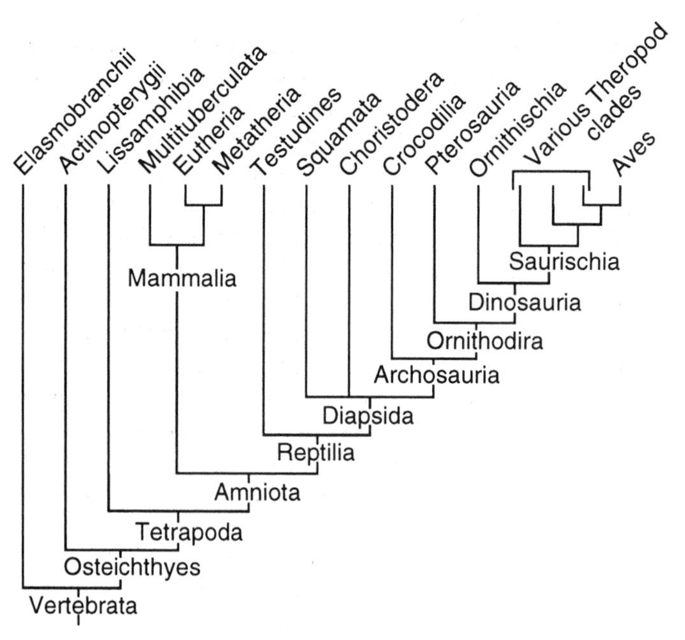

FIGURE 5.1 Phylogenetic relationships of vertebrates of the Late Cretaceous.

example, that a crocodilian is more closely related to a bird than to a lizard (a squamate). Birds and crocodilians are both archosaurs; lizards are not. But birds, crocodilians and lizards are all diapsids. At the next higher level, in which turtles (Testudines) join in, everybody is a reptile. The key to understanding this is that the emphasis must be on relationship through common ancestry.

How can we discern ancestry if we are limited to fossil and extant organisms—if, that is, we did not witness the evolution of the group in question? There is no direct approach; a cladogram can never be proved true. A cladogram is, rather, an hypothesis of relationships. Different workers may hold to different hypotheses. The basis for generating any hypothesis is an assessment of characters. Characters shared by taxa within a clade and by their presumed ancestor are termed *homologies*. For example, the extraembryonic structures discussed in chapter 2 (the amnion, chorion, and allantois) are characters found in some form in every single species of modern birds, mammals, and reptiles (and presumably dinosaurs). These characters thus almost assuredly evolved in the common ancestor of all three groups and are thus unique to these three groups.

Beyond the simple bookkeeping that this cladogram serves, it places taxa within an evolutionary framework that must undergird all questions of ecology. This framework role of a cladogram is important for understanding living creatures, but it is vastly more important for extinct forms. Two examples involving dinosaurs illustrate this point. First, although it was known that birds use gravel in digestion (and with good reason it has been argued that dinosaurs used similar, but larger stones called *gastroliths*), the strong phylogenetic implications of this digestive behavior went unrealized until the close dinosaur-bird connection was made. Second, as we began to realize that living crocodilians show some, although fairly rudimentary, care of offspring, it was suggested that dinosaurs might also have shown at least this level of care, if not the greater care found in birds. As Jack Horner's paleobehaviorial research on dinosaurs has unfolded, revealing fossil evidence of parental care of nestling dinosaurs, the prevailing cladistic hypothesis pointed to the general conclusion that dinosaurs probably had some level of parental care.

Novices using this cladogram as an easy reference for understanding relationships of the vertebrates discussed in the rest of this chapter should be aware of one feature of cladograms in general: two cladograms may look very different—and yet be saying the same thing. This is because it is the point (or using the systematist's jargon, the *node*) at which the clades unite that determines the hypothesis of relation-

Table 5.1 Survival and Extinction of Vertebrate Species Across the K/T Boundary

Species from the Upper Cretaceous Hell Creek Formation, Montana			Survivors of K/T (X)
ELASMOBRANCHII	1		
RHINOBATOIDEI			
Family indeterminate			
Myledaphus bipartitus			O
SCLERORHYNCHOIDEI			
SCLERORHYNCHIDAE			
Ischyrhiza avonicola			O
?SCLERORHYNCHIDAE			
"Squatirhina" americana			O
?ORECTOLOBIFORMES			
Family indeterminate			
"Brachaelurus" estesi			O
POLYACRODONTIDAE			
Lissodus selachos			O
NUMBER & % SURVIVAL			0/5 (0%)
ACTINOPTERYGII			
CHONDROSTEI			
ACIPENSERIDAE			
"Acipenser" albertensis			X
"Acipenser" eruciferus			X
Protoscaphirhynchus squamosus		r	O
POLYODONTIDAE			
undescribed Polyodontidae		r	X
NEOPTERYGII			
(HOLOSTEANS)			
AMIIDAE			
Kindleia fragosa			X
Melvius thomasi			O
LEPISOSTEIDAE			
Lepisosteus occidentalis			X
(TELEOSTS)			
"ASPIDORHYNCHIDAE"			
Belonostomus longirostris		r	O
Belonostomus sp.		r	X
ESOCIDAE			
Estesesox foxi	2		O
Family indeterminate			
undescribed Esocoidei	2	r	X
New, unpublished family	3		
Platacodon nanus			O
PACHYRHIZODONTOIDEI, indet.			
species indeterminate		r	O
PALAEOLABRIDAE			
Palaeolabrus montanensis		r	X
PHYLLODONTIDAE			
Phyllodus paulkatoi		r	X
NUMBER & % SURVIVAL			9/15 (60%)
LISSAMPHIBIA			
ANURA			
DISCOGLOSSIDAE			
Scotiophryne pustulosa			X
CAUDATA			
BATRACHOSAUROIDIDAE			
Opisthotriton kayi			X
Prodesmodon copei		r	X
PROSIRENIDAE			
Albanerpeton nexuosus		r	X
SCAPHERPETONTIDAE			
Lisserpeton bairdi			X
cf. *Piceoerpeton* sp.		r	X
Scapherpeton tectum			X
SIRENIDAE			
Habrosaurus dilatus			X
NUMBER & % SURVIVAL			8/8 (100%)

Species from the Upper Cretaceous Hell Creek Formation, Montana			Survivors of K/T (X)
MAMMALIA			
MULTITUBERCULATA			
CIMOLODONTIDAE			
Cimolodon nitidus			X
CIMOLOMYIDAE			
Cimolomys gracilis		r	O
Meniscoessus robustus			O
Family indeterminate			
Cimexomys minor		r	X
Essonodon browni			O
Paracimexomys priscus			O
NEOPLAGIAULACIDAE			
Mesodma formosa			X
Mesodma hensleighi			O
Mesodma thompsoni			X
? *Neoplagiaulax burgessi*		r	X
NUMBER & % SURVIVAL			5/10 (50%)
EUTHERIA			
GYPSONICTOPIDAE			
Gypsonictops illuminatus			X
PALAEORYCTIDAE			
Batodon tenuis		r	X
Cimolestes cerberoides		r	X
Cimolestes incisus		r	X
Cimolestes magnus		r	X
Cimolestes propalaeoryctes		r	X
NUMBER & % SURVIVAL			6/6 (100%)
METATHERIA			
DIDELPHODONTIDAE			
Didelphodon vorax			O
Family indeterminate			
Glasbius twitchelli			O
PEDIOMYIDAE			
Pediomys cooki		r	O
Pediomys elegans		r	O
Pediomys florencae			O
Pediomys hatcheri		r	O
Pediomys krejcii		r	O
PERADECTIDAE			
Alphadon marshi			X
Alphadon wilsoni			O
Protalphadon lulli	4	r	O
Turgidodon rhaister	5	r	O
NUMBER & % SURVIVAL			1/11 (9%)
REPTILIA			
TESTUDINES			
ADOCIDAE			
Adocus sp.			X
BAENIDAE			
Eubaena cephalica			X
Neurankylus cf. *N. eximius*	6		X
Palatobaena bairdi			X
Plesiobaena antiqua			X
Stygiochelys estesi			X
CHELYDRIDAE			
Chelydridae indet.			X
Emarginochelys cretacea			X

continued on next page

Table 5.1 Survival and Extinction of Vertebrate Species Across the K/T Boundary (continued)

Species from the Upper Cretaceous Hell Creek Formation, Montana			Survivors of K/T (X)
REPTILIA (continued)			
TESTUDINES (continued)			
KINOSTERNIDAE			
Kinosternidae indet.			X
MACROBAENIDAE			
"Clemmys" backmani			X
NANHSIUNGCHELYDIDAE	7		
Basilemys sinuosa			O
PLEUROSTERNIDAE	7		
Compsemys victa			X
TRIONYCHIDAE			
Heloplanoplia distincta			O
"Plastomenus" sp. A			X
"Plastomenus" sp. C			X
Trionyx (Aspiderетes) sp.			X
Trionyx (Trionyx) sp.		r	X
NUMBER & % SURVIVAL			15/17 (88%)
SQUAMATA			
ANGUIDAE			
Odaxosaurus piger			X
?HELODERMATIDAE			
Paraderma bogerti		r	O
NECROSAURIDAE			
Parasaniwa wyomingensis			O
SCINCIDAE			
Contogenys sloani			X
TEIIDAE			
Chamops segnis			O
Haptosphenus placodon			O
Leptochamops denticulatus			O
Peneteius aquilonius		r	O
?VARANIDAE			
Palaeosaniwa canadensis		r	O
XENOSAURIDAE			
Exostinus lancensis			X
NUMBER & % SURVIVAL			3/10 (30%)
CHORISTODERA			
CHAMPSOSAURIDAE			
Champsosaurus sp. indet.			X
NUMBER & % SURVIVAL			1/1 (100%)

Species from the Upper Cretaceous Hell Creek Formation, Montana			Survivors of K/T (X)
CROCODILIA			
ALLIGATOROIDEA	8		
Brachychampsa montana			O
undescribed alligatoroid(?) A			X
undescribed alligatoroid(?) B			X
CROCODYLIDAE			
Leidyosuchus sternbergi			X
THORACOSAURIDAE			
Thoracosaurus neocesariensis		r	X
NUMBER & % SURVIVAL			4/5 (80%)
DINOSAURIA			
ORNITHISCHIA			
ANKYLOSAURIDAE			
Ankylosaurus magniventris		r	O
CERATOPSIDAE			
Torosaurus ? latus	9	r	O
Triceratops horridus			O
HADROSAURIDAE			
Anatotitan copei	10		O
Edmontosaurus annectens			O
"HYPSILOPHODONTIDAE"			
Thescelosaurus neglectus	9		O
NODOSAURIDAE			
? *Edmontonia*sp.	9	r	O
PACHYCEPHALOSAURIDAE			
Pachycephalosaurus wyomingensis		r	O
Stegoceras validus		r	O
Stygimoloch spinifer		r	O
NUMBER & % SURVIVAL			0/10 (0%)
SAURISCHIA	11		
DROMAEOSAURIDAE			
Dromaeosaurus sp.	9		O
? *Velociraptor* sp.	9		O
ELMISAURIDAE			
? *Chirostenotes* sp.	9		O
ORNITHOMIMIDAE			
Ornithomimus sp.	9		O
TROODONTIDAE			
Paronychodon lacustris	9		O
Troodon formosus	9		O
TYRANNOSAURIDAE			
Albertosaurus lancensis	9		O
Aublysodon cf. *A. mirandus*	12	r	O
Tyrannosaurus rex			O
NUMBER & % SURVIVAL			0/9 (0%)
TOTAL NUMBER & % SURVIVAL			52/107 (49%)

Source: Updated after Archibald and Bryant 1990. Some taxa in Archibald and Bryant (1990) are excluded here because of poor taxonomy or because they are too fragmentary to confidently recognize (Dermatemydinae indet., Boidae indet., Neoplagiaulacidae gen. et sp. indet., *Paleopsephurus wilsoni, Avisaurus archibaldi, Thescelus insiliens, Ugrosaurus olsoni, Cimolestes stirtoni.)* The lower case r indicates a rare species, as defined in the text. The following species or equivalents in this table are modified from Archibald and Bryant (1990) as follows:

1 Cappetta 1987, except *B. estesi* and *"S." americana,* D. Ward pers. comm. 1994.
2 New Esocoidei named and described by Wilson et al. 1992.
3 Unpublished perciform family, Wilson and Williams 1994.
4 *Alphadon lulli = Protalphadon lulli* Cifelli 1990.
5 *Alphadon rhaister = Turgidodon rhaister* Cifelli 1990.
6 New family assignment, Gaffney and Meylan 1988.
7 New family assignment, Brinkman and Nicholls 1993.
8 Superfamily assignment after Norell et al. 1994.
9 Modifications after Weishampel et al. 1990.
10 *Anatosaurus copei = Anatotitan copei* Horner 1992.
11 Familial assignments follow Holtz 1994.
12 *Aublysodon* sp. = *A.* cf. *mirandus* Molnar and Carpenter 1989.

ships. The various clades in figure 5.1 could be rotated around their respective nodes with no change in the relationships. Thus the amniote clade could be rotated horizontally so that the mammals lie on the right and the reptiles on the left—all without changing the hypothesis of relationships in any way. A cladogram is thus like a mobile in which the various clades are free to swing around each other, while the connections remain fixed.

The particular topology that I have chosen to use in sequencing groups of organisms for a necessarily linear discussion relies on an ordering that should be familiar to most readers, but it is by no means an evolutionary ladder from the most primitive to the most derived. It should not be interpreted as sharks evolving into bony fishes into lissamphibians, and so on.

Elasmobranchs—Sharks, Skates, and Rays

The Hell Creek faunas of eastern Montana contain five known species of elasmobranchs (sharks and their relatives—all distinguished by cartilaginous skeletons). With them are found vertebrates that can reliably be gauged as exclusively or usually inhabiting fresh water or even land (table 5.1). What are elasmobranchs doing with freshwater and terrestrial vertebrates? We might be tempted, as discussed in chapter 3, to suggest some sort of reworking of older marine rocks containing elasmobranchs, but the answer comes from examining the habits of modern species.

Modern as well as fossil elasmobranchs are usually viewed as denizens of marine or brackish water. There are, however, notable exceptions. Although not occurring in any truly landlocked marine or freshwater lakes, various species of elasmobranchs are known, for example, from the Black Sea, Lake Nicaragua, Lake Izabal in Guatemala, the Ogowe River in western Africa, the Zambesi River in eastern Africa, the Tigris River near Baghdad, and the Amazon River of South America (Springer and Gold 1989). Thus some elasmobranchs are known to travel many miles upstream into fresh water.

At the time of deposition of the upper part of the Hell Creek Formation in eastern Montana, the nearest marine waters of the receding epicontinental sea were 100 to 150 miles to the east in the Dakotas. Although the cartilaginous skeletons of the elasmobranchs living in the latest Cretaceous fresh waters of eastern Montana long ago disappeared, the record of their presence is preserved in other ways—most often in the form of teeth, spines, and denticle-like scales (figure 5.2).

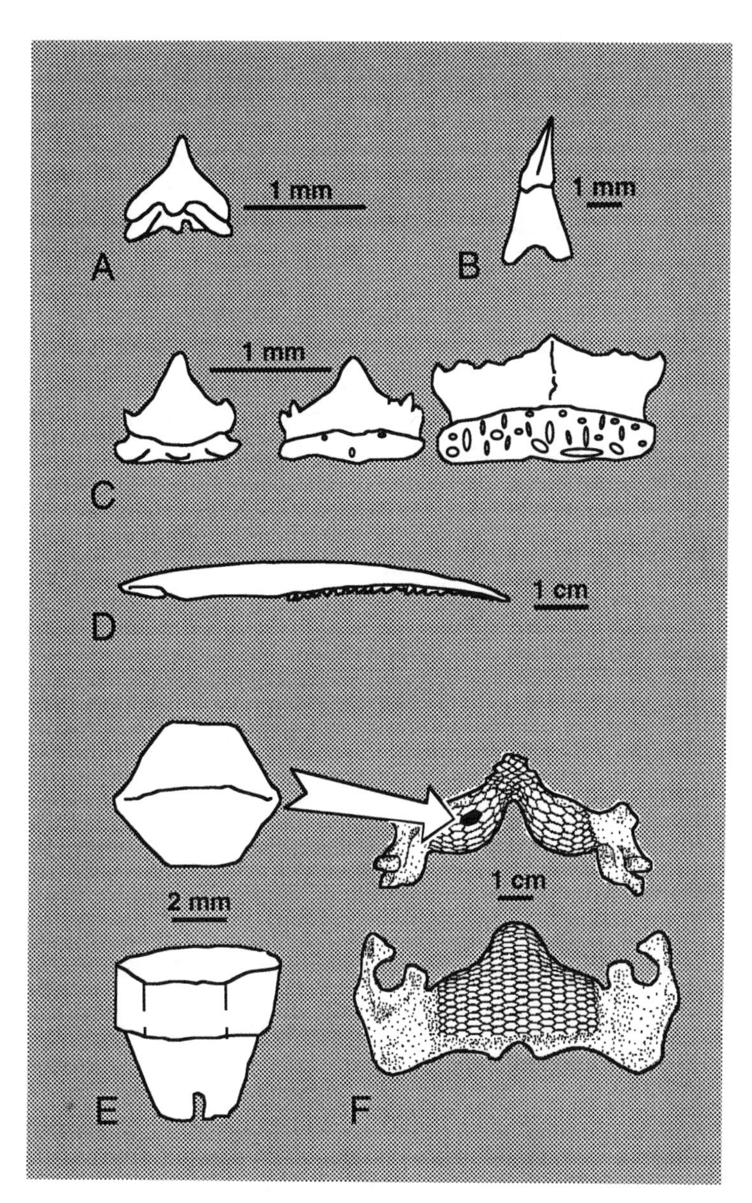

FIGURE 5.2 Elasmobranchs (sharks, skates, and rays) from the Hell Creek Formation. (A) side view of tooth of the possible sawfish *"Squatirhina" americana*; (B) front view of tooth of the sawfish *Ischyrhiza avonicola*; (C) side views of teeth (left to right) of central, slightly lateral of central, and side of jaw of the shark *Lissodus selachos*; (D) dorsal fin spine of *Lissodus selachos*; (E) top and side views of tooth of the ray *Myledaphus bipartitus*; (F) jaws of the modern dasyatid ray *Hylophus sephen*, showing pavement teeth similar to isolated teeth of *Myledaphus bipartitus*. *Source: A, B, C, E, and F after Estes 1964; D after Bryant 1989.*

Because the kind of fossil evidence for elasmobranchs is limited, assignment to families can be difficult. At least three and possibly as many as five family-level groups of elasmobranchs are present in the Hell Creek fauna. The polyacrodontids are represented in the Hell Creek fauna by a single species, *Lissodus selachos* (C and D of figure 5.2), that may well represent one of the last of this family, which first appears in the rock record as long ago as the Early Triassic almost 250 mya (Cappetta 1987). One and possibly two sawfishes are part of the Hell Creek fauna: *Ischyrhiza avonicola* (figure 5.2B) and *"Squatirhina" americana* (figure 5.2A). Some extant sawfishes frequent fresh water (Nelson 1984). A ray, *Myledaphus*, is common in the Hell Creek fauna (figure 5.2E). Some modern rays frequent fresh water.

By drawing parallels with the ecologies of modern elasmobranchs, we have good reason to conclude that the Hell Creek species made their livings by feeding off or near the bottom, sometimes actively rooting in the sediments to find prey.

Actinopterygii—Bony Ray-Finned Fishes

Today, bony fishes—more correctly, the actinopterygians, or ray-finned fishes—include more than 20,000 species. This is almost half of the modern diversity of backboned animals (Moyle and Cech 1988).

In the Late Cretaceous faunas of Montana, many of the modern groups of ray-finned fishes had not yet evolved or were only beginning to diversify. Thus, of the fifteen species in the Hell Creek fauna, almost half represent more ancient lineages, all of which still survive today, though some with far fewer species. Two of these more ancient lineages are both chondrosteans: the acipenserids (sturgeons) and the polyodontids (paddlefish).

As the term chondrostean suggests, these fishes have an almost completely cartilaginous skeleton, although sturgeons do have large bony scutes and spines (figure 5.3A). From what can be discerned from the fossil material, which ranges in quality from miscellaneous scutes to very rare nearly complete specimens, these fish are strikingly similar to their modern counterparts. Sturgeons feed by probing the bottom with their snouts. As they grow, their diet shifts from invertebrates to fish. Sturgeons are the largest of living freshwater fish, reaching a length of almost twenty-eight feet, although the largest adults are not strictly freshwater inhabitants. Paddlefish, today restricted to the Yangtze River drainage in China and the Mississippi River system, are filter feeders that capture plankton by swimming with their mouths agape (Moyle and Cech 1988). Although the Late Cretaceous

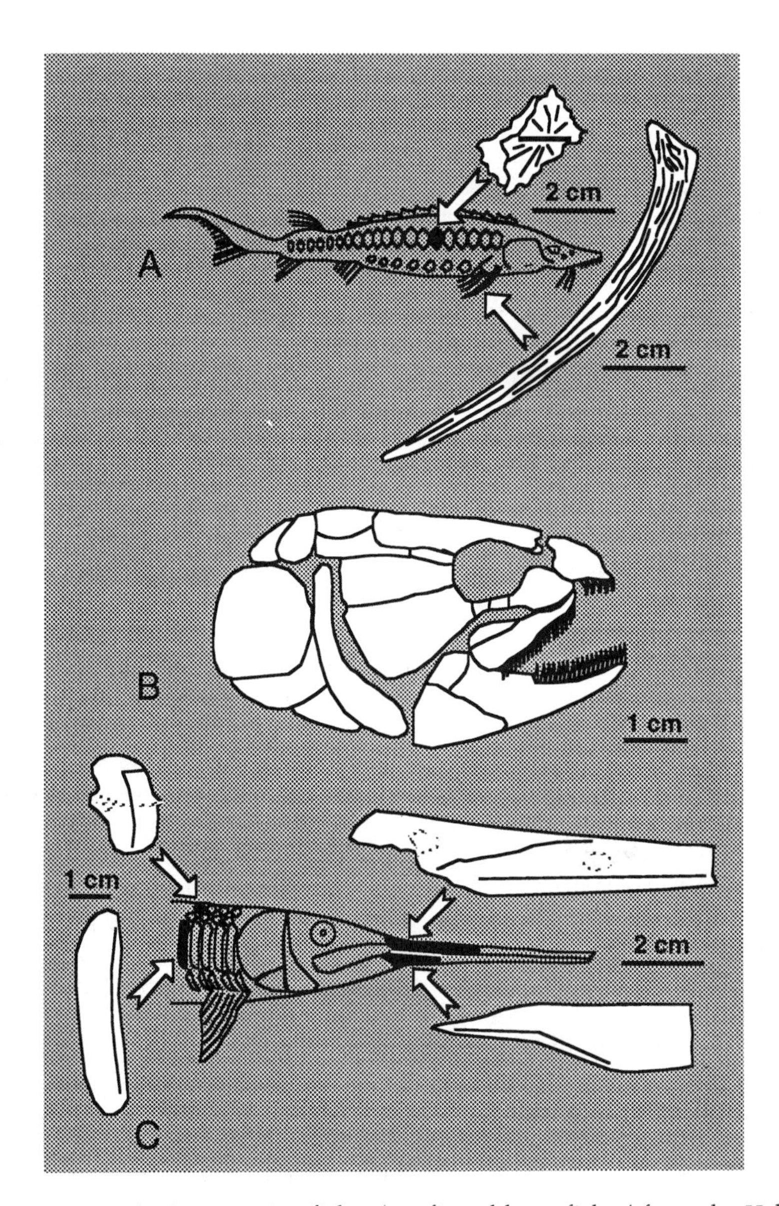

FIGURE 5.3 Actinopterygian fishes (ray-finned bony fishes) from the Hell Creek Formation. All are right side views. (A) a bony scale and a pectoral fin spine from the sturgeon *"Acipenser" eruciferus*, with positions based on the modern sturgeon *Acipenser* sp.; (B) restored skull of the bowfin *Kindleia fragosa*; (C) two kinds of bony scales and upper and lower jaw fragments of a teleost of uncertain affinities, *Belonostomus longirostris*. *Source: A after Estes 1964 and Carroll 1988; B and C after Estes 1964.*

forms also seem to have been filter feeders, some early Tertiary relatives may not have fed in a similar fashion.

By far the greatest diversity of ray-finned fishes is today found in the Neopterygii. In addition to the bony fishes with which most people are familiar, this group also includes two lineages that retain many primitive characteristics of the group. These two lineages are the lepisosteids (gars) and the amiids (bowfins) (figure 5.3B). Both are voracious predators, mostly on other bony fishes. Gars are especially well known as lay-and-wait predators that suddenly dart out, snatching their prey by the tail. Gars are among the most heavily armored living fishes. The weight of this armor is compensated by a large swim bladder that helps them maintain neutral buoyancy. The swim bladder in gars as well as in bowfins allows these fish to breath air, although neither is an obligatory air breather. The distributions of extant gars and bowfins is centered on the Mississippi River drainage (Moyle and Cech 1988).

Together, sturgeon, paddlefish, bowfin, and gar account for seven of the fifteen ray-finned fishes in the Hell Creek fauna. What is rather remarkable about this pattern is that all of these groups of fish are still found in the Mississippi River drainage, the descendant of the rivers that drained the Late Cretaceous landscape of North America.

The remaining eight neopterygians are teleosts, the group that today overwhelming includes the greatest number of fish species. Five of these remaining species are allocated to four extinct families, while two are aligned with extant families. According to the Canadian paleoichthylogist Mark Wilson (pers. comm.; see Wilson and Williams 1994), the eighth species, *Platacodon nanus*, belongs to a new, as yet unpublished, extinct family of perciform fishes (the group including perch).

An extinct family of teleosts, the aspidorhynchids, are cosmopolitan, predominantly marine fishes of the Jurassic and Cretaceous (Carroll 1988), although they are now known to survive well into the Paleocene in freshwater deposits in North America (Bryant 1989). With their elongate snout they superficially resemble gars, but features of the skull show they belong among the teleosts (figure 5.3C).

Pachyrhizontoids are limited to the Cretaceous, but they appear to have had a nearly worldwide distribution. In Late Cretaceous strata, they are moderately common fishes in marine deposits of the epicontinental seaways of mid-America. Past systematic work aligned them with elopiforms, the extant order of fishes that includes tarpons, but Forey (1977) moved them to a more uncertain position among the teleosts.

Bryant (1989) discussed the uncertainties regarding the affinities of palaeolabrids. Although most evidence points toward their being teleosts, better material suggests closer affinities with gars.

As with palaeolabrids, the relationship of the next family, Phyllodontidae, is also not clear, although they have been implicated in the ancestry of eels (Carroll 1988).

Finally, Mark Wilson and his colleagues recently reexamined Late Cretaceous and early Tertiary freshwater fish faunas that earlier had been described by Estes (1964) and Bryant (1989). They reported (Wilson et al. 1992) that among some of the very small teleost bones, some of which had been incorrectly assigned to *Platacodon*, were bones of the extant lineage Esocoidei. This is the lineage including escids (pike) and the much smaller umbrids (mudminnows). In the Hell Creek fauna they could recognize two species. Although much smaller than extant representatives, one of the new species, *Estesesox foxi*, belonged with the pike. Other tiny specimens were suggestive of the mudminnows, but were too fragmentary for a more specific identification. Although very different in size, pike and mudminnows are both voracious lay-and-wait predators, albeit on different prey. It will be especially interesting for zoogeographers if it proves true that there are umbrids in the Late Cretaceous faunas, because they and pike are obligatorily freshwater fishes.

Unlike today, the Hell Creek freshwater fish fauna is not dominated by teleosts. There are two possibilities for this pattern. From what we can tell of teleost evolution, this group only began to replace other ray-finned fishes in the Cretaceous. The perciforms are among several teleost groups that commenced their tremendous radiation only at the very end of the Mesozoic. Even today in the Mississippi River drainage, the flavor of the Hell Creek fish fauna remains with the presence of paddlefish, sturgeon, gar, and bowfin. Another possibility for the dearth of teleosts was suggested by Wilson and his colleagues (Wilson et al. 1992). They observed that the Late Cretaceous record is rather fragmentary compared to that for the Tertiary. Museum collections contain numerous Late Cretaceous teleostean fossils that have yet to be examined. Thus, teleosts may yet prove to be much more numerous in the Late Cretaceous than now thought.

Lissamphibia—Frogs and Salamanders

Lissamphibia, meaning "smooth double life," is an apt term for this nonamniote group of tetrapods. These scaleless, hairless, featherless tetrapods usually split their life between an exclusively aquatic larval stage and an adult stage that may or may not take place on land. But reproduction must take place in water or a very wet setting. Fossils are unlikely to preserve scales, hair, feathers, and reproductive strategies. Therefore, we distinguish Lissamphibia from amniote tetrapods in the fossil record by features of their skeleton.

Lissamphibia is a monophyletic group that includes three major extant lineages—frogs and toads, salamanders and newts, and the legless caecilians. The more encompassing Amphibia is not a monophyletic group and thus for clarity and consistency is not to be used (you will not find it in figure 5.1). It includes more ancient Paleozoic relatives that comprise the ancestry of not only lissamphibians but also amniote vertebrates—reptiles (including birds) and mammals.

The presence of seven species of salamanders and one species of frog in the Hell Creek faunas is one of the best indicators of permanent bodies of fresh water (figure 5.4). Nevertheless, lissamphibians are not as common in the Hell Creek faunas as they are to the south in the comparably aged Lance fauna in eastern Wyoming. The Lance Formation, which has produced the Lance fauna, is dominated by large fluvial deposits. While lissamphibians are an important component of the Hell Creek fauna, a more even distribution of environments is represented in the Hell Creek than in the Lance, and thus lissamphibians are relatively less common.

The presence of several species or a large number of individuals of lissamphibians in the rock record is taken to indicate that the aquatic setting, whether riverine or lacustrine, is fresh water. This remains the most conservative, but not always best, assessment. First, lissamphibian remains could be swept out into marine waters, there to be buried among marine species. I have found such material, as well as teeth of terrestrial mammals, in nearshore marine deposits. Second, lissamphibians, though rarely, have been reported in salty or brackish water; Neill (1958) attributes these tendencies to eleven extant species of salamanders and thirty of frogs and toads. Only three of the salamanders but twenty-six frogs or toads clearly frequented or bred in brackish to salt water. These findings indicate that we must use some caution and ancillary evidence in determining whether particular deposits are fresh, brackish, or salty. In the case of the Hell Creek fauna, all seven species of salamander—except for possibly a batrachosauroidid salamander, *Prodesmodon*—appear to have been mostly aquatic (Bryant 1989). Overall, it is very likely that the Hell Creek aquatic faunas lived in fresh, rather than marine, waters.

Of the five families of lissamphibians from the Hell Creek, only one, Sirenidae, includes living species. The other four families persist

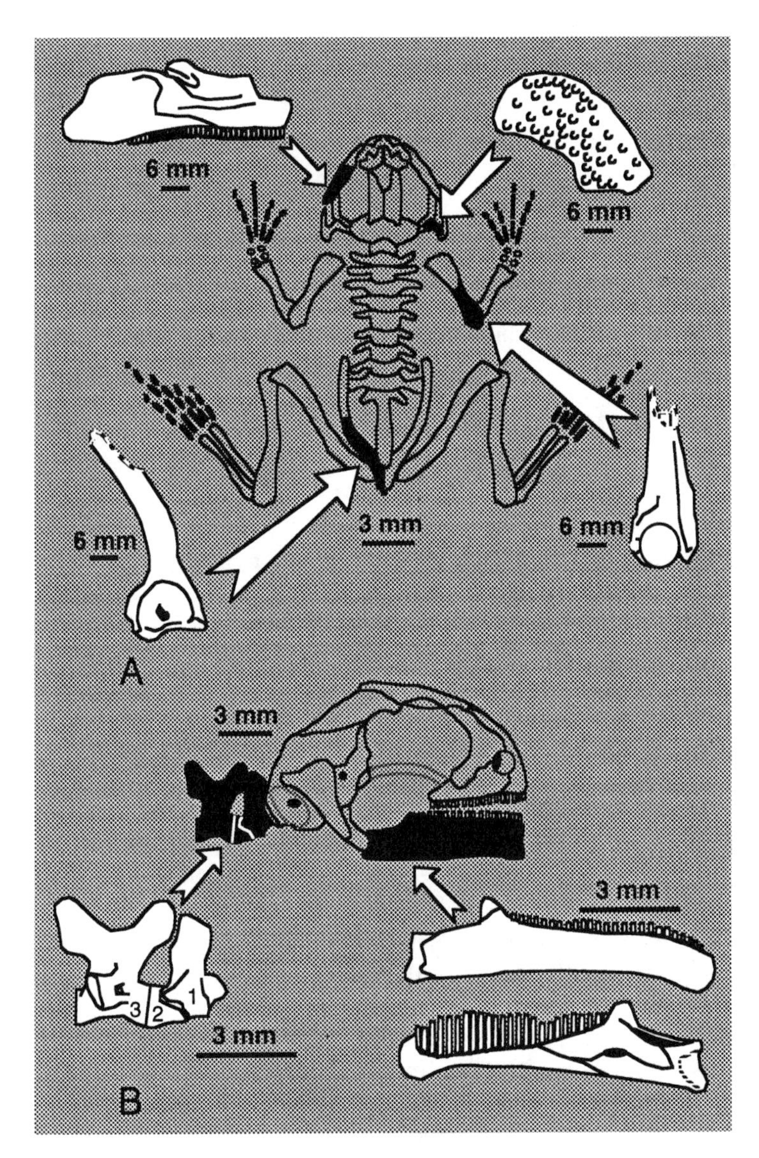

FIGURE 5.4 Frog and salamander from the Hell Creek Formation. (A) *clockwise from upper right:* right skull bone fragment (squamosal), right lower humerus, left pelvic (ilial) fragment, and left upper jaw (maxilla) of the discoglossid frog *Scotiophryne pustulosa*, all positioned within a generalized frog skeleton;(B) right side views of first through third vertebrae and right lower jaw (above), and inside view of right lower jaw (below) of the unusual salamander *Albanherpeton nexuosus*, and as positioned within a better-known Miocene species, *Albanherpeton inexpectatum. Source: A after Estes 1969b; B after Estes 1982 and unpublished data.*

into the Tertiary, but most show their heyday in the earlier part of the Tertiary. All five families make their first appearance in the Late Cretaceous, but in faunas older than the Hell Creek fauna.

Mammalia—Multituberculates, Placentals, and Marsupials

Mammals inherited the terrestrial realm for larger vertebrates after the extinction of dinosaurs. During the reign of the dinosaurs, however, mammals were small, mostly nocturnal or crepuscular creatures skulking in the underbrush or in the overarching canopy. The very largest among the mammals were no bigger than an overfed raccoon. I hope I may be forgiven for waxing somewhat more personally about these beasts. This the group on which I do most of my research.

By the demise of the dinosaurs 65 mya, the two great lineages of living mammals had been established: the marsupials (or metatherians) and the eutherians (or placentals), which together are called therians. Placentals, which derive their name from the specialized extraembryonic membrane that sustains the growing embryo, are the lineage to which we belong. Of course, no fossil evidence directly informs us that these early placentals indeed had a placenta, so I will refer to them by their other sobriquet—eutherians, which means "true beasts." Similarly, many but not all living marsupials possess a marsupium, and this characteristic too is not fossilizable; so I will refer to them as metatherians, which means "near beasts." As fossils, eutherians can be distinguished as usually having only three molars, while most metatherians have four (compare figures 5.6 and 5.7).

The split between metatherians and eutherians stretches back at least a hundred million years (still in the Cretaceous). Although North America may seem an unlikely cradle for the metatherians, the earliest fossil of this group was described by Rich Cifelli (1993b) from deposits more than a hundred million years old in central Utah. For this earliest marsupial he chose a rather playful name, *Kokopellia*, after the enigmatic hunchbacked flute player depicted in Anasazi rock art. For those of us who have worked in the American southwest where petroglyphic representations of Kokopelli abound, this was a particularly delightful epithet for the earliest metatherian. The earliest accepted eutherians come from localities in the Gobi Desert of Mongolia and the deserts of western Asia in the former Soviet Union. The placental material from the Gobi has been most thoroughly and recently described by the renowned Polish paleomammalogist Zofia Kielan-Jaworowska and various colleagues (e.g., Kielan-Jaworowska and Dashzeveg 1989); the middle Asian material was discovered and described by Lev Nessov (e.g., Nessov 1987). Because metatherians and eutherians are nearest relatives (or "sister groups," as we say in systematics), eutherian origins must also date to at least 100 mya and probably closer to 110 mya. This is Rich Cifelli's contention, presented in his most recent overview of metatherian evolution (1993a).

The third major lineage of mammals noted in table 5.1 and figure 5.1 is the extinct Multituberculata. The record for the group extends to at least 150 mya. Multituberculates are easily distinguished from other mammals by the distinctive morphology of their teeth; the name multituberculate refers to the multiple tubercles, or cusps, aligned in rows in the back two teeth. They are grouped as mammals because of their skull and skeletal features. The affinities of this distinctive lineage are still controversial. Traditional opinion puts the origin of multituberculates (among the first mammals) back to almost 200 mya—solidly in the Late Triassic. More recent opinion, such as that of Tim Rowe (1993), places them as the nearest sister group to the therians. Even more recently, Zofia Kielan-Jaworowska and Petr Gambaryan (1995) argue that the supposedly close evolutionary ties between multituberculates and therians are not supported, based upon their study of the postcrania of multituberculates. This would indicate that multituberculates are a very ancient and separate mammalian lineage.

Although there is much we do not know or understand about Late Cretaceous mammal evolution, mammals are arguably the best-studied vertebrates within this interval, although dinosaurs are closing rank fast. Undoubtedly the best preserved mammalian fossils of this age come from the Gobi Desert, where in some cases complete skulls and skeletons have been collected. The importance of the Gobi to mammal evolution first came to light in the 1920s, when the American Museum of Natural History, then directed by the flamboyant Roy Chapman Andrews, launched expeditions to this remote region. Political exigencies put an end to the American effort in Mongolia. In the 1940s and 1950s Soviet paleontologists worked in the Gobi, as did the Poles in the 1960s and 1970s. Both groups collaborated with their Mongolian colleagues.

Although all teams have come away with treasure troves of fossil vertebrates, the Poles were particularly attentive to bringing their findings to the attention of the scientific community, in the form of publications. This is especially true for the mammals, largely through the drive of the leader of the Polish expedition, Zofia Kielan-Jaworowska. Today, the American Museum of Natural History, as well as a host of other institutions and workers from the United

States, Europe, and Asia, are once again discovering vertebrate treasures in the Gobi Desert.

As I discussed in chapter 1, for a number of reasons—good roads, good outcrops and fossils, and more paleontologists—North America has the best-studied sequence of vertebrates (especially mammals) through the last thirty million years of the Cretaceous leading up to the K/T boundary. What these fossils lack in quality, compared to the Gobi Desert specimens, they more than make up for in quantity. This allows a much tighter sequencing of vertebrate faunas in North America than is yet possible in the Gobi.

In the Hell Creek fauna, multituberculates and metatherians are of similar taxonomic diversity, with ten and eleven species, respectively. Eutherians are represented by just six species. In the latest Cretaceous of Alberta, where fewer species of multituberculates and marsupials have been discovered, the three lineages are more similar in diversity. Nevertheless, the proportions of species in the Hell Creek fauna seem to be representative of the latest Cretaceous in western North America (Archibald 1982).

Although multituberculates are superficially similar to rodents, with their enlarged pair of upper and lower incisors (figure 5.5), multituberculates were probably more omnivorous, eating not only plants but also invertebrates, especially arthropods. Of course many living rodents also supplement their diet with insects. If the North American opossum, *Didelphis virginiana*, is any indication, most of the Hell Creek marsupials were probably quite catholic in their tastes (figure 5.6A and B). An exception may have been *Didelphodon vorax* (figure 5.6C); besides being one of the largest mammals in the fauna, it had bulbous premolars—possibly used in crushing bone—that resemble those of the modern Tasmanian devil, which consumes the carcass of its prey (or discovered carrion), hair, bone, and all. Most of the Hell Creek eutherians probably were strictly insectivores or carnivores. This is suggested by the very high, sharp crests and cusps on the teeth, useful for slicing and dicing insect carapaces (figure 5.7).

Testudines—Turtles

Apart from fish scales, fragments of turtles are among the most commonly found in the Hell Creek faunas. Being aquatic certainly enhances their likelihood of preservation, but as Howard Hutchison and I showed (1986), the Hell Creek turtle fauna ranks among the richest known—either fossil or recent. Seventeen species in the Hell Creek represent at least eight different families of turtles (table 5.1).

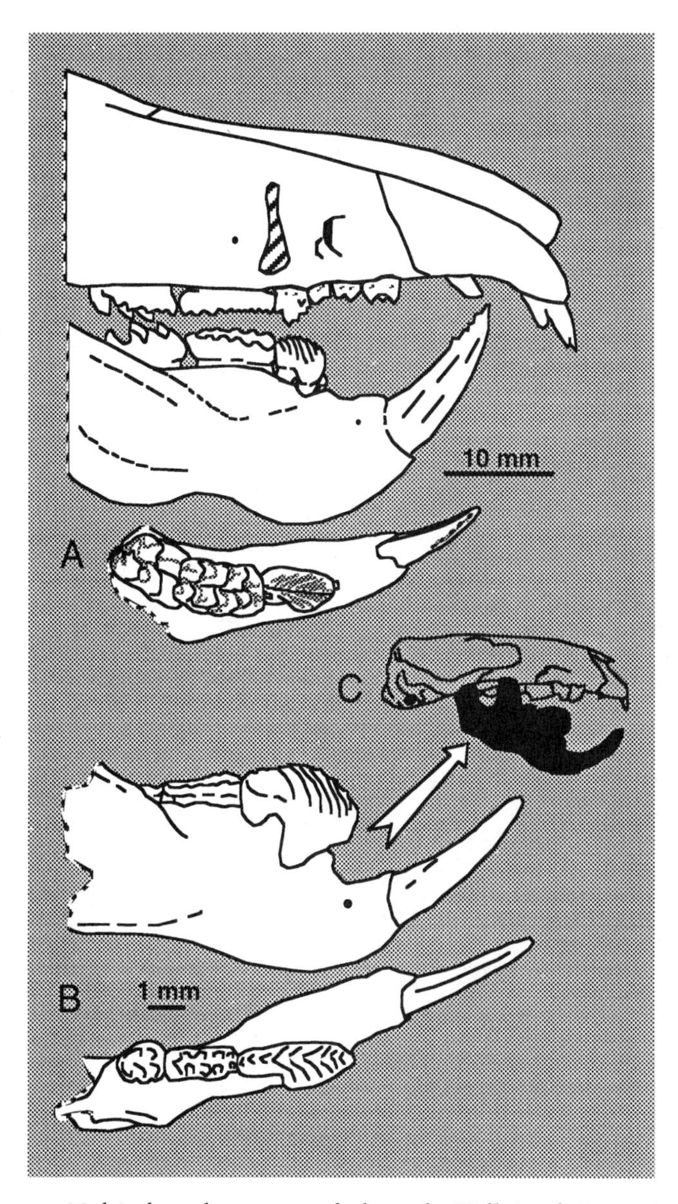

FIGURE 5.5 Multituberculate mammals from the Hell Creek Formation. (A) right side view of restored snout and lower jaw (above) of *Meniscoessus robustus* (top view of lower jaw shown below); (B) right side (above) and top views (below) of lower jaw of *Mesodma formosa* (side and top views); (C) outline of right side of largely hypothetical skull and lower jaws of a ptilodontoid multituberculate, the group to which *Mesodma* belongs. *Source: A after Archibald 1982; B and C after Clemens and Kielan-Jaworowska 1979.*

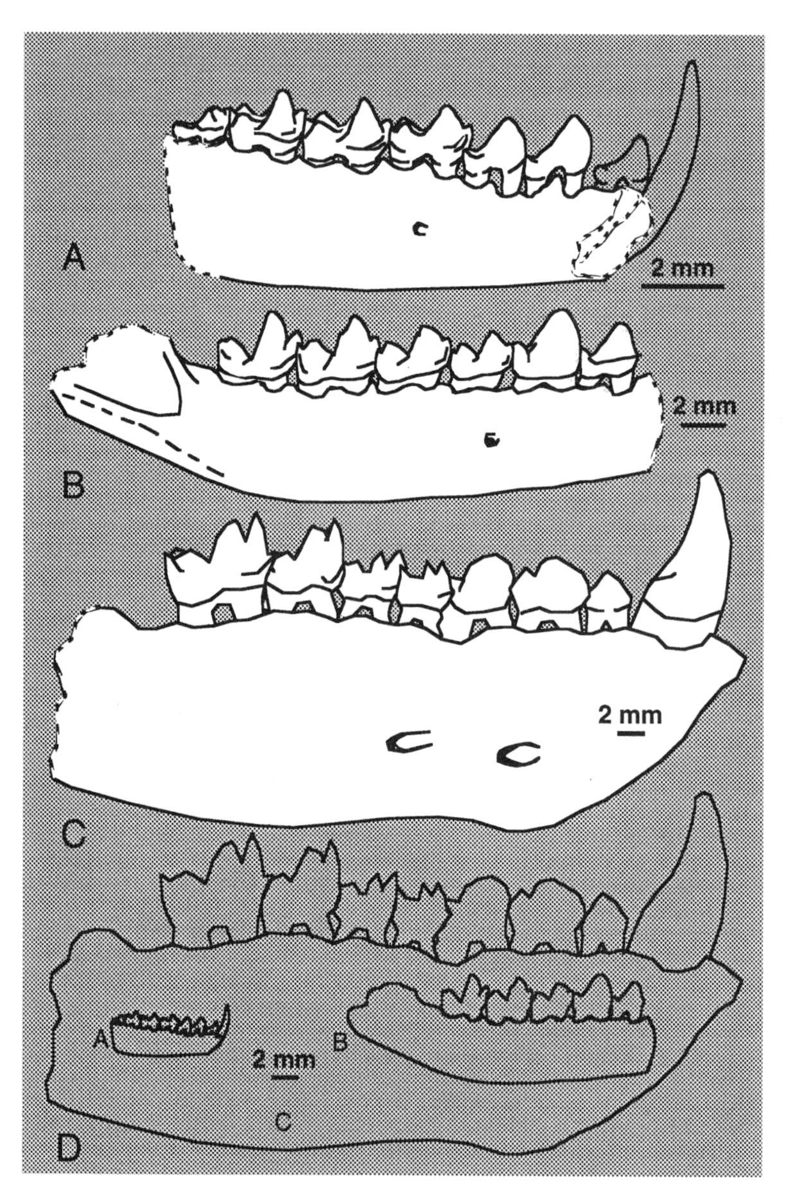

FIGURE 5.6 Metatherian (marsupial) mammals from the Hell Creek and Lance formations. All are right side views. (A) restored lower jaw (outline of canine and first premolar, remaining two premolars, all four molars) of *Glasbius intricatus*, a marsupial of uncertain familial affinities from the Lance Formation, Wyoming, which is better known than the closely related Hell Creek Formation species *G. twitchelli*; (B) restored lower jaw (two of three premolars and all four molars) of the pediomyid *Pediomys hatcheri*; (C) restored lower jaw (canine, all three premolars, all four molars) of the stagodontid *Didelphodon vorax*; (D) outlines of the above three metatherian lower jaws shown at the same scale, which is about twice natural size. *Source: A and B after Clemens 1966; C after Clemens 1968.*

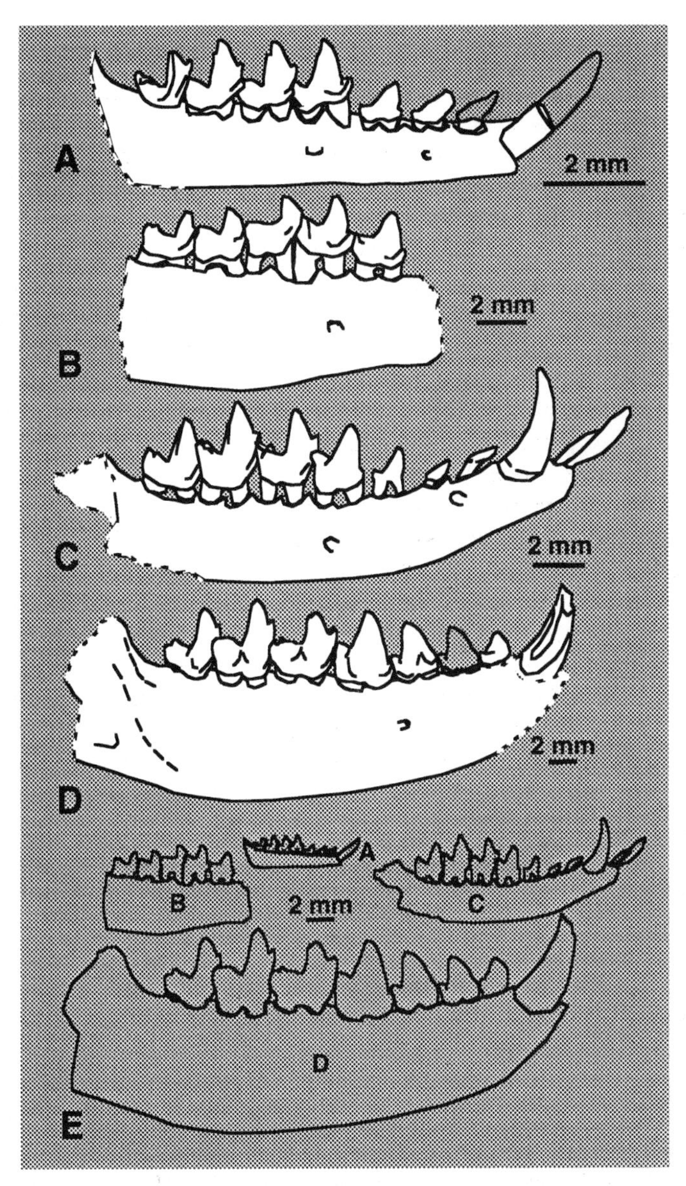

FIGURE 5.7 Eutherian (placental) mammals from the Hell Creek Formation. All are right side views. (A) restored lower jaw (partial canine, outline of first premolar, remaining three premolars, all three molars) of the palaeoryctid *Batodon tenuis*; (B) partially restored lower jaw (two of five premolars, all three molars) of the gypsonictopid *Gypsonictops illuminatus*; (C) restored lower jaw (two of three? incisors, canine, all four premolars, all three molars) of the palaeoryctid *Cimolestes propalaeoryctes*; (D) restored lower jaw (canine, three of four premolars, three molars) of the palaeoryctid *Cimolestes magnus*; (D) outlines of the above four eutherian lower jaws shown at the same scale, which is about twice natural size. *Source: A after Lillegraven 1969 and Clemens 1973; B, C, and D after Lillegraven 1969.*

The turtle fauna includes an interesting mix of extinct lineages, especially the Baenidae, which is at its evolutionary acme near the K/T boundary, and the first inklings of at least three modern families, including snapping turtles (chelydrids) (figure 5.8A and E). A fourth extant family, the "soft-shelled" turtles (trionychids), showed considerable diversity in the Hell Creek (figure 5.8F).

Because turtles lack teeth, the mouth is modified into a beaklike structure covered with cornified tissue that provides a hard cutting edge and crushing surface. The shape of the cutting edge and the angle and size of the crushing surfaces are suggestive of diet. For example, among baenids, the skull and jaws range from narrower and longer in *Plesiobaena* (figure 5.8C) to shorter and broader in *Palatobaena* (figure 5.8D), the latter probably used in crushing snails and freshwater clams that are abundant in the same deposits.

The only turtle in the assemblage that is suggestive of a terrestrial existence is *Basilemys*. It is fairly rare in the Hell Creek fauna probably because it may have been mostly terrestrial. When present, it is easily recognizable by the chicken wire–like pattern on the shell. Its stubby elephantine feet, well-developed limb armor, and heavily constructed lower shell (plastron) with a prominent knob on the front end (figure 5.8H) are features often shared by today's terrestrial tortoises. Although clearly not a true tortoise—based upon its shell characteristics—*Basilemys* nevertheless anticipates the features of extant tortoises. The shell may have reached a length of three feet, a formidable size.

Aquatic modifications are obvious in the baenids, an extinct family. One curious feature of the baenids, however, is shared by only a very few modern turtles. Apparently, the individually growing bony plates that form the mosaic patterns of each shell fused upon reaching adult size. Thereafter the shell no longer expanded; growth (especially of the lower plastron) was instead invested in shell thickening. The thickness of the lower plastron approaches an inch and a half in some specimens. Any possible selective reason for this growth pattern remains unknown. The slightly domed and relatively fusiform shell, extremely long tail, and recurved claws all suggest a strong, bottom-walking turtle. Added to these anatomical observations is the fact that well over half of the shells were recovered from channel sandstones, indicating moving water.

The single most common family of turtles alive today is Trionychidae. The common name for this group, "soft-shell," refers to the flexible margins of the shell resulting from the loss of bony elements that encircle the shells of other turtles (figure 5.8F).

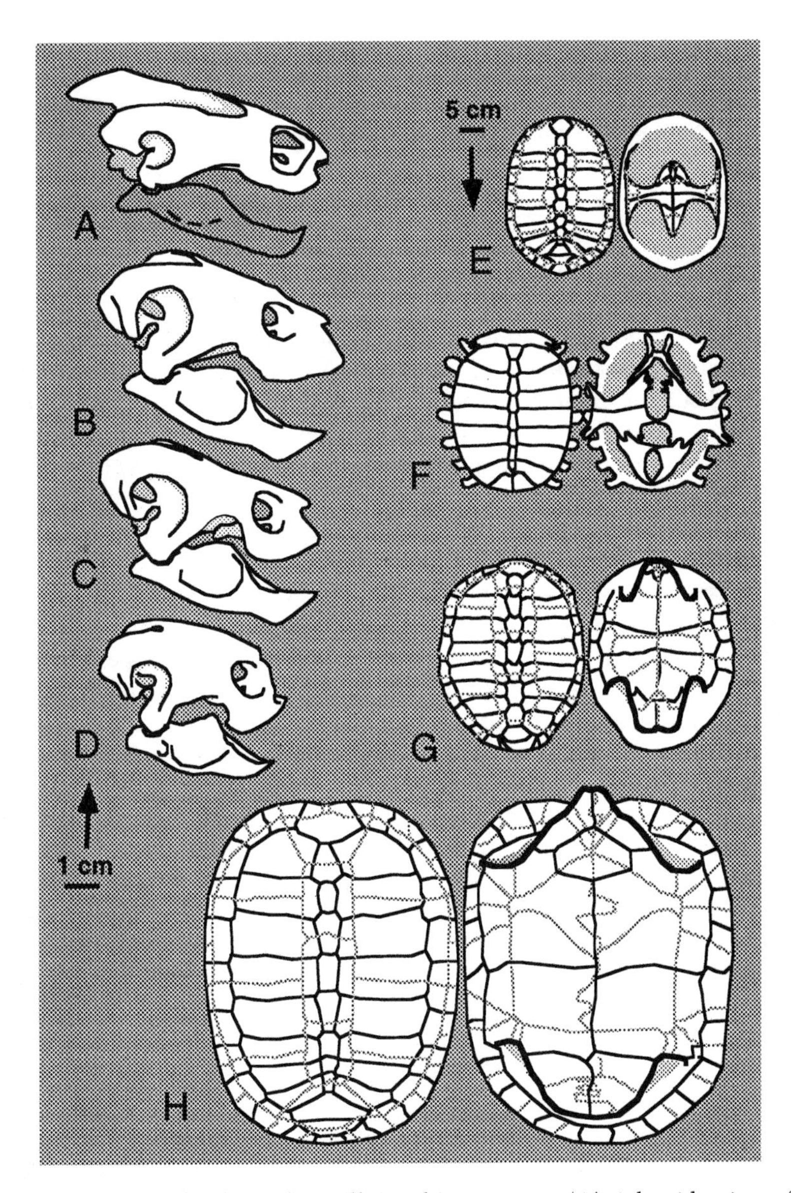

FIGURE 5.8 Turtles from the Hell Creek Formation. (A) right side view of restored skull and outline of lower jaw of the snapping turtle *Emarginochelys cretacea*; (B) right side view of restored skull and lower jaw of the baenid turtle *Eubaena cephalica*; (C) right side view of restored skull and lower jaw of the baenid turtle *Plesiobaena antiqua*; (D) right side view of restored skull and lower jaw of the baenid turtle *Palatobaena bairdi*. In E through H the top, or carapacial, part of the shell is on the left and the bottom, or plastral, part is on the right. Sutures between the bones are black, while margins between the horny scutes are dotted. (E) *Emarginochelys cretacea*; (F) the soft-shelled turtle *Trionyx* (*Aspideretes*) sp.; (G) *Plesiobaena antiqua*; (H) the somewhat tortoise-like *Basilemys sinuosa*. A through D are at the same scale, as are E through H. *Source: A and E after Whetsone 1978; B, C, G after Gaffney 1972 and Archibald 1977; D after Archibald and Hutchison 1979; F and H after Hay 1908.*

Identification and relationships among trionychids remain obscure, but five or six species may have frequented the latest Cretaceous waterways in eastern Montana. Presumably, these species, like their extant counterparts, were exclusively carnivores.

Squamates—Lizards and Snakes

One of the "gee whiz" bits of information of which many people are not aware is that snakes are merely a kind of lizard, albeit legless. The lizard group has branched into forms that lost or reduced their legs a minimum of eight times. One of those excursions produced the squamate group that we call snakes. Snakes seem to have been the most successful; there are considerably more species of snake than other legless or nearly legless forms. Snakes had certainly arisen by latest Cretaceous times, but they had not yet undergone their considerable radiation. Vertebrae belonging to one unknown species of boid (the family to which boa constrictors belong) have been recovered from the Hell Creek fauna, but this species is so rare that we can dispense with snakes in this review and move on to the more numerous lizards. Because the lizards are more common, we can be much more confident that their fossil pattern of extinction and survival across the K/T boundary in eastern Montana is more accurate.

The ten species of Hell Creek lizards in table 5.1 are assigned to seven families. Of these, only the Necrosauridae is extinct today. Of the six extant families, all but Varanidae (monitor lizards) are still found in North America, although the greatest diversity for the other families today is subtropical to tropical. Among the tetrapods in the Hell Creek fauna, lizards show the greatest diversity in adult body size, ranging from about six inches in the smaller teiids (figure 5.9C and D) to more than ten feet for the possible varanid, *Palaeosaniwa* (figure 5.9E).

As authors such as Estes (1964) and Bryant (1989) discuss, some of these lizard families have a wide range of ecological preferences, so using the lizards to circumscribe aspects of ecology in the latest Cretaceous is not advisable. From what is known, most of the lizards should be categorized as terrestrial. A semiaquatic habitus cannot be precluded, however; the teiids even today are know to include semiaquatic species. As to diet, again the little that is known suggests insectivory through carnivory (figure 5.9A & E), depending upon size. There is a hint of distinct variety in lizard diets of the Late Cretaceous, such as the heavy, crushing teeth of *Odaxosaurus piger* (figure 5.9B), which Richard Estes (1964) found suggestive of a diet of heavy-shelled arthropods or mollusks.

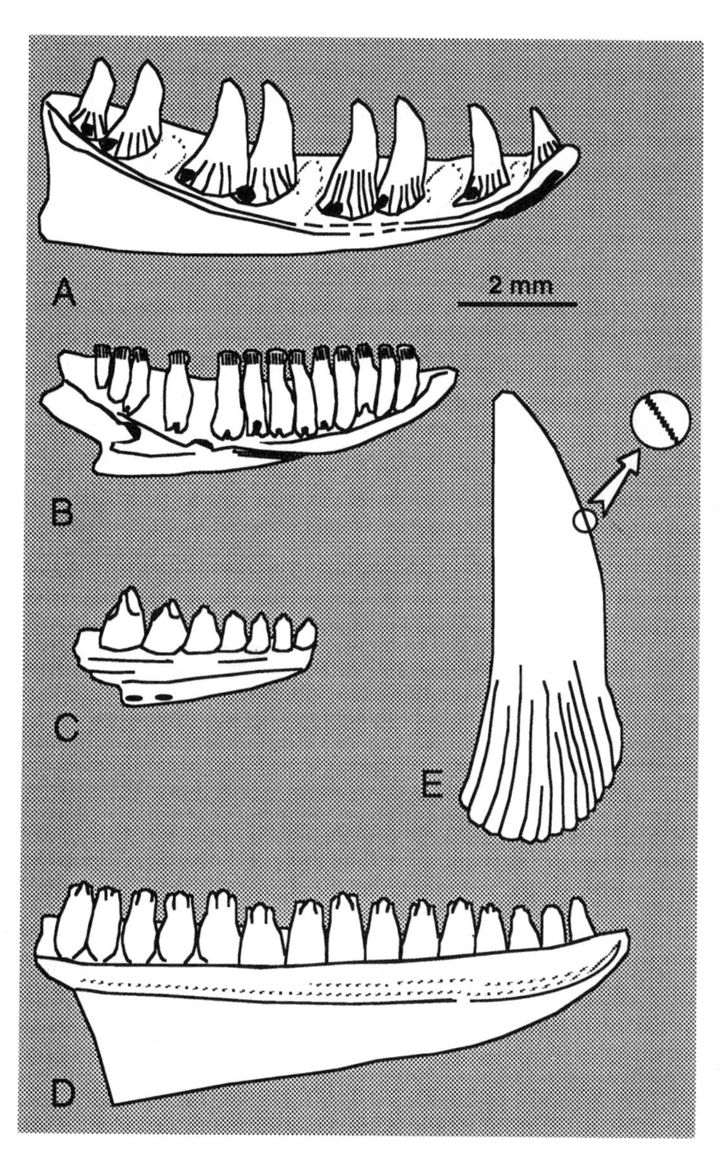

FIGURE 5.9 Lizards from the Hell Creek Formation. The inside, or tongue side, view of four lower left jaws and one left tooth (at the same scale) of (A) the necrosaurid *Parasaniwa wyomingensis*; (B) the anguid *Odaxosaurus piger*; (C) the teiid *Penetius aquilonius*; (D) the teiid *Chamops segnis*; (E) the possible varanid *Palaeosaniwa canadensis*. *Source: A, B, D, E after Estes 1964; C after Estes 1969a.*

Given the apparent diversity of at least size in the Hell Creek lizards, the high levels of extinction at the K/T boundary (seven of ten species) call for some explanation. In chapter 8, I will suggest one possibility for this pattern.

Choristoderans—Champsosaurs

Of all the major groups of vertebrates in the Hell Creek faunas, probably the least familiar is Choristodera, or champsosaurs, as the North American forms are more commonly known. Champsosaurs first appear in the Late Cretaceous of North America. Except for possible Early Cretaceous relatives, we have few clues as to their origin.

At a glance, these creatures seem to be a form of narrow-snouted crocodilian, such as the extant gavials of Asia. On closer inspection of the skull and skeleton, however, champsosaurs are clearly not crocodilians. Characters of the skull, notably the two openings behind each eye, show that champsosaurs belong to the larger group of reptiles including lizards, snakes, crocodiles, pterosaurs, and dinosaurs (including birds) known as Diapsida. The name itself refers to the two openings, or more correctly the bony arches, behind the bony eye socket (figures 5.10A, C, and E). As the cladogram in figure 5.1 shows, however, champsosaurs cannot be aligned with either the group including lizards and snakes—the Squamata (Lepidosauria in Evans 1988)—or the group including the crocodilians, pterosaurs, and dinosaurs (including birds)—the Archosauria (Gauthier et al. 1988). Thus, in the cladogram, Squamata, Choristodera, and Archosauria form a three-way split, suggesting this uncertainty. The latest Cretaceous and early Paleocene champsosaurs are known from very well preserved material, but this does not alleviate the problem of uncertain relationship because these forms are already so specialized that it is difficult to interpret their anatomy within a phylogenetic context.

Although complete skulls and skeletons are quite rare, fragmentary champsosaur remains are often found. Champsosaurs may well have been very common animals—but not necessarily. Because their aquatic environments are preferentially preserved and because parts of their skeleton, such as their vertebrae, are so distinctive, they may appear in the fossil record to be relatively more common than they may have been in life.

Bruce Erickson described some of the best-preserved material of champsosaurs (1972), found in North Dakota. Although his was Paleocene material, very few differences can be discerned between

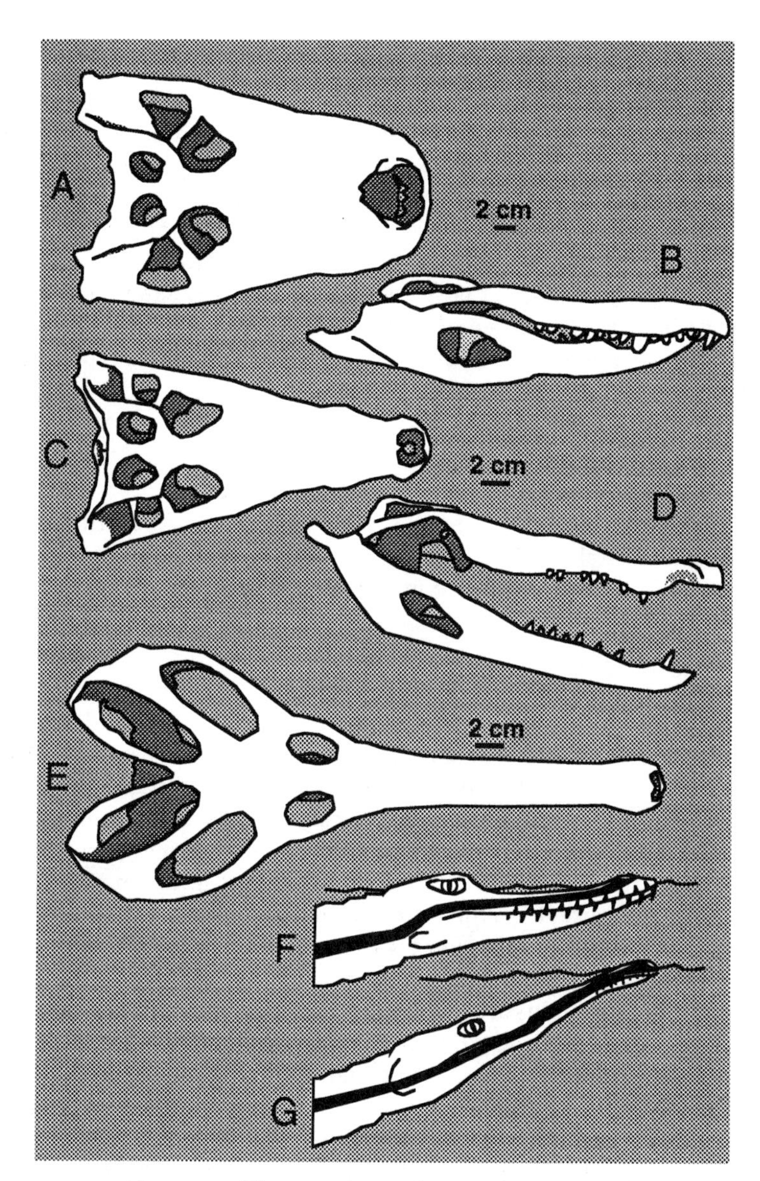

FIGURE 5.10 Two crocodilians and one choristoderan from the Hell Creek Formation. (A) top and (B) right-side views of the skull of the alligator *Brachychampsa montana*; (C) top and (D) right-side views of restored skull of the crocodile *Leidyosuchus sternbergi*; (E) top view of restored skull of the choristoderan *Champsosaurus* sp.; (F) crocodilian at water surface, showing both exposed eye and snout (no scale); (G) choristoderan at water surface, showing only exposed snout (no scale). *Source: A and B after Norell et al. 1994; C and D after Gilmore 1910; E, F, G after Erickson 1972.*

material from the latest Cretaceous and the Paleocene. Skeletal reconstructions suggest that champsosaurs held their limbs close to the body and swam with strong flexions of the body. Erickson points out that unlike crocodilians that can bring both eyes and narial openings just above the water to stalk prey (figure 5.10F), champsosaurs' eyes remained well below water as the tip of the snout broke the surface to breathe (figure 5.10G). This suggests that champsosaurs may have relied on a more strictly aquatic diet.

Species-level identifications remain problematic for champsosaurs, so none is suggested in table 5.1.

A SURVEY OF THE ARCHOSAURS

The rest of this chapter reviews what is known about the latest Cretaceous complement of archosaurs—that is, crocodilians, ornithischians, saurischians, and the pterosaurs. Appropriately, archosaur means "ruling reptile."

Crocodilia—Crocodiles and Alligators

Contrary to popular knowledge, crocodilians are not simply overgrown aquatic lizards. From both fossil evidence and studies of modern vertebrates, crocodilians and birds share a more recent common ancestor with each other than either does with lizards—or with any other living reptile, for that matter. This relationship is shown in the cladogram in figure 5.1. When the extinct dinosaurs are included, the crocodilians are seen to be somewhat more distant to birds, but nevertheless closer to them than to lizards.

Modern crocodilians are of three lineages (families): crocodiles, alligators, and gavials. Crocodiles are usually distinguished from alligators by their more pointed snout and teeth exposed in side view. Gavials are easy to distinguish because of their thin snout. In the Hell Creek fauna, two of these lineages—the crocodiles and the alligators—are discernible. The heavily pitted fragments of skull bone have the stamp of a crocodilian that distinguishes them from other archosaurs. The more infrequent skull and jaw material provides assignment to the clades that include modern families.

The business end of the skull in *Leidyosuchus sternbergi* shows, among other things, the kind of narrow snout found in most but not all crocodiles alive today (figure 5.10C and D). The teeth in this species are also similar to modern crocodiles; they have tall, pointed crowns. The broad, rounded snout of *Brachychampsa montana* likewise is suggestive of the snout in many modern alligators (figure 5.10A and B). In its dental morphology, however, it is different from modern alligators in having short, blunt teeth, suggestive of an animal using its teeth for crushing turtle shells. (Recall, aquatic turtles are exceedingly common in this fauna.) Bryant (1989) briefly described additional jaws, bony scutes, and other fragments found by Howard Hutchison that are highly suggestive of two other kinds of alligators in the Hell Creek and younger faunas. A few fragments have also been recovered from the Hell Creek that resemble the marine crocodile *Thoracosaurus* known from the Eocene of New Jersey; little else can be said about this material.

Ornithischia—Bird-Hipped Dinosaurs

In the Hell Creek Formation, the ornithischian branch of Dinosauria is represented by ten species in six families—all of which failed to cross the K/T boundary into the Tertiary (table 5.1). They are all medium to very large terrestrial herbivores. The only other (presumed) large herbivore of the Hell Creek is a tortoiselike turtle, *Basilemys*.

The ornithischians are commonly known as the bird-hipped dinosaurs because part of their pubic bone is bent backward, superficially resembling the hip in birds. This distinguishes them from the saurischians, the so-called lizard-hipped dinosaurs, whose pubic bones retain the primitive, unbent condition. But, as I warned in an earlier chapter, don't let the name confuse you. Today's birds are more closely related to the lizard-hipped dinosaurs than to the bird-hipped dinosaurs.

Ornithischian dinosaurs ranged from human-size, for the bone-headed pachycephalosaurid *Stegoceras* (figure 5.11C), up to hippo- or elephant-size for hadrosaurids (duck-bills; figure 5.11A) and ceratopsids (horned dinosaurs; figure 5.11D). Weight is somewhat harder to estimate than length, so we cannot be sure if *Stegoceras* was the counterpart of a tackle or safety for a pro football team. Estimates for weight vary from author to author, but in a delightful book on the dynamics of dinosaurs, the functional anatomist McNeill Alexander (1989) gave a range of weights for *Triceratops*, drawing from various sources. This ornithischian was pegged at 6.1 to 9.4 tons. Estimates for *Anatotitan* (*Anatosaurus* in Alexander) ranged from 3.4 to 4.0 tons. Alexander gave estimates of about 2.5 tons for a male hippo and 5.5 tons for a bull African elephant.

Ornithischians never rivaled in taxonomic diversity the extant complement of medium to very large mammalian herbivores. What they lacked in numbers of species they may well have compensated for with large herds (Horner and Gorman 1988).

Although it might be tempting to carry the analogy with large mammalian herbivores further, caution is warranted. When we think of large-mammal herbivore faunas, places such as the Rift Valley of Africa come to mind. Most of the plants, especially the grasses that these mammals are munching, had not, however, evolved or diversified even by latest Cretaceous time. Thus, these ornithischian herbivores were dining on other kinds of plants. Possibly the reason for the lower number of species of these dinosaurs, as compared to counterparts among mammals, has more to do with the variety of plants upon which they could feed than anything intrinsic to dinosaurs.

Dentitions strongly point to herbivory in all the ornithischians in the Hell Creek fauna, but beyond this, only one other general observation is possible. For most dinosaurs, the evidence is lacking or ambivalent concerning the processing of food in the mouth—that is, chewing—whether we are dealing with carnivores or herbivores. Hell Creek ornithischians such as the pachycephalosaurids, the ankylosaurs (both ankylosaurids and nodosaurids), and hypsilophodontids had relatively few, small teeth (compared to comparably sized mammalian herbivores), suggesting that most food processing was in the gut and little in the mouth. The two notable ornithischian—and, in fact, dinosaur—exceptions are among ceratopsids and hadrosaurids. Both these major lineages possessed batteries of teeth that wore and were shed and replaced. The elegant studies of Weishampel (e.g., 1983) on hadrosaurids and Ostrom (1966) on ceratopsids have shown that replacement and chewing are very different in these two lineages. Nevertheless, both were certainly capable of processing in their mouths even the toughest plant material.

Except for the hadrosaurids, there seems to be general agreement that the Hell Creek ornithischians were all terrestrial. For a time, paleontologists portrayed them as fully or semiaquatic, in part because hadrosaurids have been found in marine and nearshore sediments, because they have "duck-bills," and because of supposed impressions of webbed front feet. The pendulum seems to have swung to a more terrestrial interpretation, but foraging may not have been exclusively on land (Weishampel and Horner 1990).

As I noted in chapter 3, the number of kinds of dinosaurs declined during the last ten million years of the Cretaceous, from the Judith River fauna to the Hell Creek fauna. A conservative estimate is a 33% decline, but it could have been as high as 45%, depending upon whose

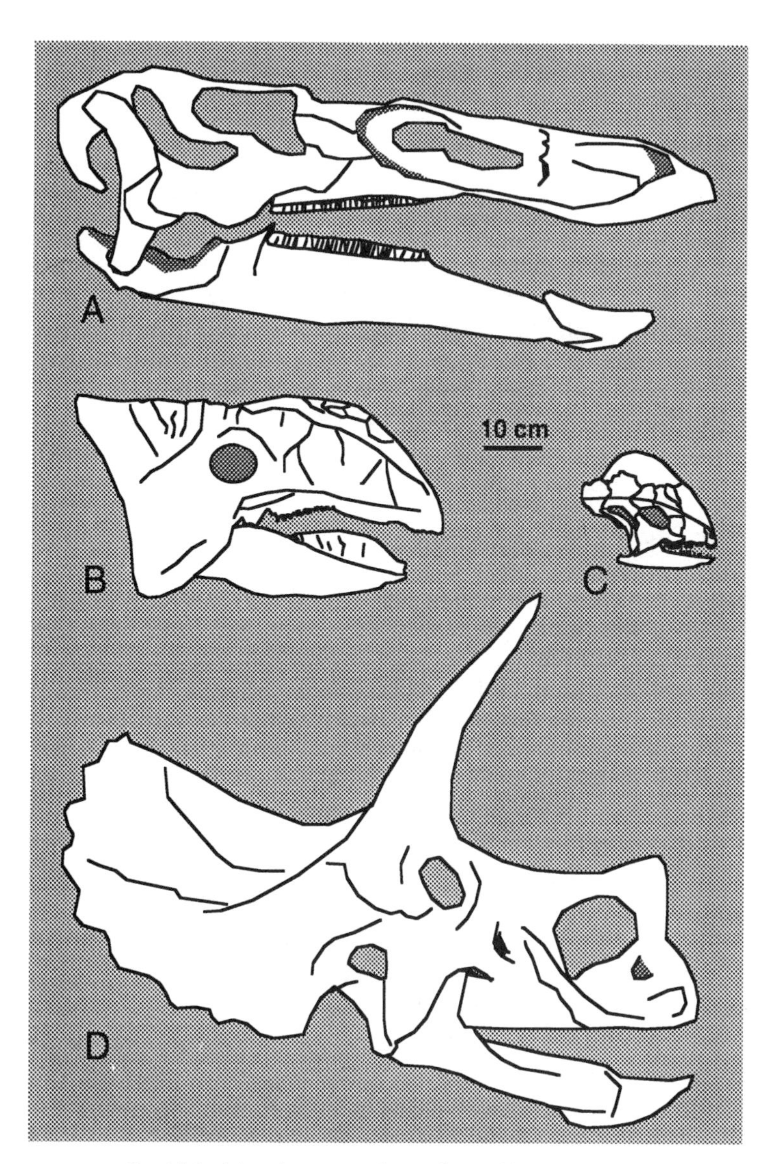

FIGURE 5.11 Ornithischian dinosaurs from the Hell Creek Formation. (A) the hadrosaurid *Anatotitan copei*; (B) the ankylosaurid *Ankylosaurus magniventris*; (C) the pachycephalosaurid *Stegoceras validus*; (D) the ceratopsid *Triceratops horridus*. All skulls and jaws are restored right side views and at same scale. *Source: A after Weishampel and Horner 1990; B after Coombs 1978; C after Maryn'ska 1990; D after Dodson and Currie 1990.*

taxonomy is used and whether species or higher taxa (such as genera) are used. Most of this decline is within Ornithischia, especially among hadrosaurids and ceratopsids. The number of genera of just these two families goes from twelve in the Judith River fauna to only four in the Hell Creek fauna. Although we cannot be sure how many species this generic decline represents, it is reasonable to extrapolate that it was of the same magnitude. Thus, in these two families alone the decline is 66%. No special pleading can obscure this very real pattern of decline. If a comparable percentage of medium to large mammalian herbivores disappeared over a comparable period of time during the Tertiary, we would certainly try to explain it rather than try to explain it away.

Saurischia—Reptile-Hipped Dinosaurs

The reptile-hipped dinosaurs, or saurischians, include two major clades: the sauropodomorphs and the theropods. The sauropodomorphs include the really big guys, such as *Diplodocus* and *Brachiosaurus* of the Jurassic. Saurischians from the Hell Creek fauna are all theropods. There is no question that sauropodomorphs are present in various other Late Cretaceous dinosaur faunas. I have even collected their peglike teeth from such localities. For some unknown reason, however, they are absent from Late Cretaceous dinosaur faunas in the northern part of the Western Interior. To my knowledge, the furthest north that any sauropodomorph bones have been found in the latest Cretaceous is southwestern Wyoming.

Some of the species of theropods in the Hell Creek fauna are represented only by extremely fragmentary specimens—though these fragments may be moderately common, such as *Paronychodon*, which is known in the Hell Creek only from teeth. From what we can glean, *Paronychodon* had a mouth replete with teeth and was probably carnivorous. With the exception of a questionable varanid, *Palaeosaniwa*, which was presumably a ten-foot lizard similar to the extant Komodo dragon, all the medium to very large carnivores in the Hell Creek fauna were theropods. They ranged from the human-size *Troodon* (figure 5.12D) to the very large *Tyrannosaurus rex* (figure 5.12E).

The three largest theropods—*Tyrannosaurus*, *Albertosaurus*, and *Aublysodon* (figure 5.12A)—are all tyrannosaurids, according to a recent phylogenetic study by Thomas Holtz (1994). *T. rex* has no size match among at least living terrestrial carnivores. The largest, the brown bear, tops out at slightly over three-quarters of a ton. Alexander (1989) provides estimates of from 4.5 to 7.7 tons for *T. rex*. The general

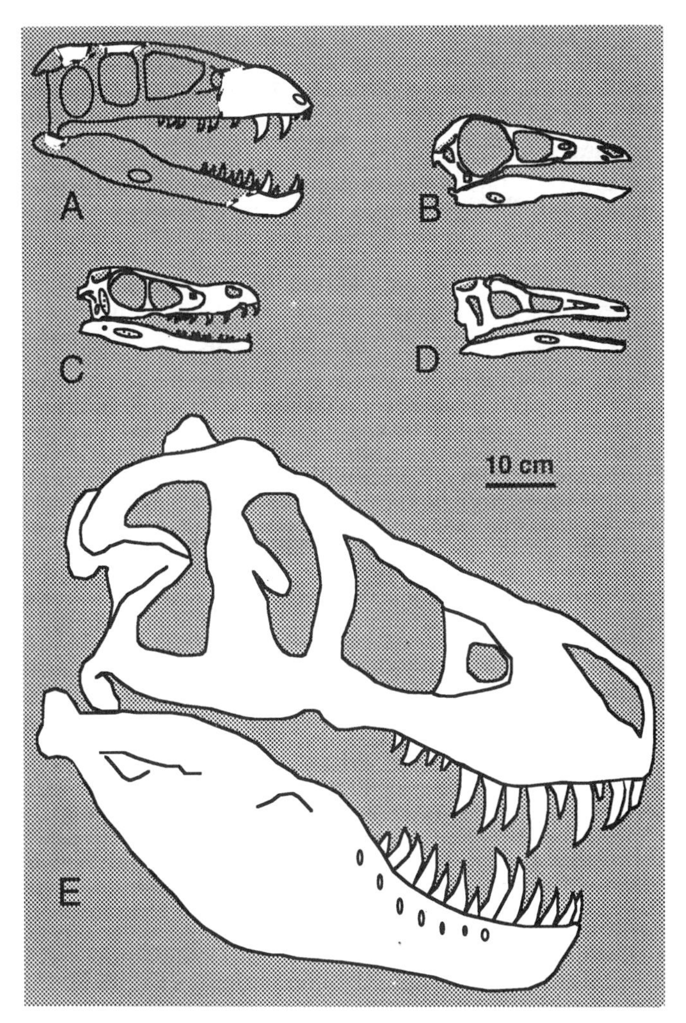

FIGURE 5.12 Saurischian dinosaurs from the Hell Creek Formation. (A) the smaller tyrannosaurid *Aublysodon* cf. *A. mirandus*; (B) the ornithomimid *Ornithomimus* sp. (restored after *Struthiomimus*); (C) the dromaeosaurid *Veliceraptor* sp.; (D) the troodontid *Troodon formosus* (restored after *Saurornithoides*); (E) the largest tyrannosaurid, *Tyrannosaurus rex*. All skulls and jaws are restored right side views and at same scale. *Source: all are after Molnar and Carpenter 1989; D is also after Osmólska and Barsbold 1990.*

perception of *T. rex* is as a blood-thirsty (and lawyer-eating, according to *Jurassic Park*) predator. No one doubts *T. rex* was a carnivore, but whether it was an active predator capturing its own prey or a scavenger using its formidable stature to secure the best pickings of the dead and dying is now a matter of controversy (Horner and Lessem 1993).

Next down in the scale of size is *Albertosaurus*, which was a slightly more gracile—that is, slightly more lightly built—tyrannosaurid, weighing in at two tons (Lambert 1990). Finally, the smallest of the tyrannosaurids is *Aublysodon* (figure 5.12A). Molnar and Carpenter (1989) suggest a rather gracile form from that skull. Here is a theropod that would have been about the weight of an adult male lion, perhaps 450 pounds.

The remaining six species of saurischian all appear to be smaller, more agile theropods, roughly the size of a six-foot human—although some, such as *Dromaeosaurus*, were larger, and others, such as *Velociraptor* (figure 5.12C) and *Troodon* were slightly smaller. In the cases where we know something about the skeleton of the animal, they had elongate forelimbs, often bearing long claws that would have been of considerable aid in the capture of prey. All but one, *Ornithomimus*, were also armed with a battery of cutting teeth, suggesting carnivory. The presence of large pedal claws in dromaeosaurids makes active predation almost a given. The diet of *Ornithomimus* (figure 5.12B), with its toothless, beaked mouth, has been of some debate (e.g., Norman 1985; Osmólska and Barsbold 1990). With its long and flexible hands and arms, and elongate hindlimbs for speed, it could easily have chased down small swift prey, such as lizards and mammals, but could also have taken advantage of softer plant material.

I have indicated (Archibald 1989b) that the fossil record shows that saurischians, like ornithischians, declined from the Judith River fauna some 75 mya to the Hell Creek fauna about 65 mya. The decline of about 25% for saurischians was not as dramatic as the decline among the hadrosaurids and ceratopsids. Even though the actual number of species drops into the Hell Creek, some saurischians may have been enjoying a continuing appearance of new forms.

The Fliers—Aves and Pterosauria

Although birds and pterosaurs are part of the Hell Creek fauna, they are so rare that to include them in any analysis would present a very skewed portrait of their contribution to the fauna.

Laurie Bryant noted in her 1989 monograph on nondinosaurian lower vertebrates that only about thirty specimens of birds were reported from the combined record of both the Hell Creek Formation and the overlying Paleocene Tullock Formation. She suggested that these thirty specimens may represent only four different species of birds. A fifth can now be added to this list. In 1985 Michael Brett-Surman and Gregory Paul named a new dinosaur *Avisaurus archibaldi*, based on part of an ankle (tarsometatarsus) that I had discovered. While all agree that it is a theropod in the broad sense (remember, birds are theropods), there was no consensus as to whether *A. archibaldi* was more closely related to the more typical ground-dwelling theropods (i.e., a birdlike theropod such as *Ornithomimus*) or whether it had truly crossed the avian Rubicon (i.e., a theropodlike bird such as *Archaeopteryx*). In 1992 a paleornithologist, Luis Chiappe, convincingly argued that *A. archibaldi* was on the avian side of the divide. In 1993 evidence for this assignation became overwhelming. Howard Hutchison found and reported a partial skeleton of *Avisaurus* that included all the hallmarks of a flighted bird, including fully developed wings. Although found in the Kaiparowits Formation in southern Utah, which is about ten million years older than the Hell Creek Formation, the feet of the two specimens are nearly identical, save for the smaller size of the Kaiparowits specimen.

Part of the early confusion as to whether *Avisaurus* fell on the nonavian or avian side of the dinosaur clade stemmed from the peculiar way in which the bones of the ankle fuse. In modern birds the three ankle bones (more correctly, metatarsal bones) fuse from the bottom up, while in the most ancient birds they fuse from the top down. *Avisaurus*, along with other birds showing the same kind of top-down metatarsal fusion, has been placed in a group of wholly Mesozoic birds known as enantiornithines, or "opposite birds" (Chiappe 1992; Feduccia 1995). It turns out that this downward fusion of the metatarsals is actually a primitive retention in enantiornithines, as it is also found in the earliest bird, *Archaeopteryx*, plus several lineages of nonavian theropod dinosaurs (Chiappe, pers. comm., 1995)—just one more indication that birds are indeed dinosaurs.

In an essay published in *Science* in 1995, Alan Feduccia indicated, especially in his accompanying phylogeny, that the enantiornithines were undergoing a major radiation that was cut off at the K/T boundary. Further, the modern lineage of birds did not appear until after the K/T boundary. Feduccia pointed out that he was not advocating a catastrophic or a gradual extinction, only that the events also had a dramatic effect on birds. From my considerable work on Late Cretaceous vertebrate faunas, I seriously doubt that we can as yet make any pronouncement about whether birds did or did not undergo a dramatic turnover at the K/T boundary. There simply is no fossil record to argue one way or the other.

The patterns of extinction of old, and origination of new, lineages of birds expressed by Feduccia is not shared by all paleornithologists. One such person is Luis Chiappe, who has published a number of scientific papers on Mesozoic birds. He is now preparing several more general, popular accounts (Chiappe 1995). Chiappe agrees with Feduccia that enantiornithines do disappear by the end of the Cretaceous, but he points out that he knows of only five species of Maastrichtian enantiornithines from around the world (pers. comm. 1995). Further, it appears that only *Avisaurus archibaldi* from the Hell Creek Formation can confidently be regarded as late Maastrichtian. The impression given by Feduccia of a radiation of such birds up to the K/T boundary is thus highly misleading. Chiappe further believes (pers. comm. 1995) that the evidence shows at least three lineages of modern birds in the Late Cretaceous (the lineage that includes ducks and relatives, the loon clade, and the shorebird clade). The presence of these three modern bird lineages in the Late Cretaceous also necessitates that each of their nearest related clades, as well as the earliest splitting clade of modern birds (the ground-dwelling ratites, such as ostriches), were also present in the Late Cretaceous. This means that more than three modern bird clades were well on the way to establishing themselves before the K/T boundary.

Pterosaurs, a distinct clade of nondinosaurian archosaurs, are exceedingly rare in the latest Cretaceous deposits around the world. Is this rarity an artifact of poor preservation or does it signify something else? The bones of birds and pterosaurs are less sturdy than those of other archosaurs because they are very thin-walled. In 1964 Richard Estes noted some pterosaur bones including what he identified as a very large tail vertebra in the latest Cretaceous Lance fauna from eastern Wyoming. He indicated its similarity to *Pteranodon*, which is at best three-quarters the size of the later described giant *Quetzalcoatlus*. He puzzled over the large size of the vertebra. Lev Nessov, who named the pterosaur family Azhdarchidae, realized that this was a neck vertebra and that its large size suggested it could belong to a giant azhdarchid such as *Quetzalcoatlus* (Nessov 1984).

In his exquisitely illustrated book on pterosaurs, Peter Wellenhofer (1991) notes only the Azhdarchidae extending up to the K/T boundary. The huge azhdarchid *Quetzalcoatlus* may have been the only pterosaur to make it to the K/T boundary in the Western Interior. Kevin Padian, another world expert on pterosaurs, affirms that all the latest Cretaceous faunas include only azhdarchids (pers. comm. 1994). According to the fossil evidence, therefore, pterosaurs may have been dwindling well before the Cretaceous ended, but it must be remembered just how fragile the bones of pterosaurs are. The fossil evidence, in this case, is suggestive—but far from conclusive.

VI

The Fates of Vertebrates Across the K/T Boundary

Without question there are more theories of how dinosaurs became extinct than for any other creatures that have inhabited this planet. In the late 1980s Alan Charig, curator emeritus of the British Museum of Natural History, informed me that he had tallied over eighty theories of how dinosaurs had met their demise. As I will argue in the closing chapter, it was probably not a single cause that brought the dinosaur extinction, but eighty causes surely is overkill!

I will not dwell on those theories for which physical or biological evidence is lacking, but will go directly to the three that have been most recently tested and debated. These are the *impact theory*, the *volcanism theory*, and the *marine regression theory*. I regard all of these as ultimate, not proximate. By this I mean that none of these events would have directly caused the extinctions. Rather each would have precipitated ancillary or corollary consequences that then did the deed.

This distinction is important. Moreover, it may be time for the debaters to join forces. Although doubters as to an impact, or volcanism, or marine regression near or at the K/T boundary remain in force, evidence is becoming overwhelming that all three occurred. In chapters 7 and 8 I will spend some time discussing these three events, especially magnitude and timing, but I accept them all as having been adequately demonstrated. What most requires testing, in my view, is whether none, some, or all of the corollaries to these theories happened and what their biological consequences may have been. These are the proximate causes of extinction.

A number of approaches can be taken for examining the corollaries, but I will largely restrict myself to testing them within the framework of the terrestrial vertebrates and their ecological settings. As I noted in the prologue, my approach is unabashedly from a vertebrate perspective—in large measure because this is my area of expertise, but also because books on the subject of the K/T boundary have almost wholly ignored the vertebrate record and what it says about theories of extinction. I leave it to others to someday undertake the prodigious work of synthesizing the implications of all the fossil faunas and floras across the K/T boundary.

PREVIOUS STUDIES

In the last chapter I introduced the cast of characters—the vertebrate species from the Hell Creek Formation in eastern Montana. This formation is currently the best known for studying events near the K/T boundary, but it certainly is not the only or earliest discovered formation yielding latest Cretaceous vertebrates. Discoveries of latest Cretaceous vertebrates in the Western Interior of North America began in the latter part of the nineteenth century. U.S. government surveying parties were the early finders. Many new species of vertebrates were named at that time, especially by the dueling giants of paleontology of the last century, Othniel Charles Marsh and Edward Drinker Cope.

Not until the middle of this century, however, did researchers undertake the essential task of assessing this amalgam of vertebrate species as a complete fauna. In the mid 1960s, publications on mammals by Bill Clemens (1964, 1966, 1973) and lower vertebrates by Richard Estes (1964) began to provide a much more thorough picture of the latest Cretaceous vertebrate fauna. The fossils in these studies were recovered from the uppermost Cretaceous Lance Formation in eastern Wyoming. In the 1960s work began in earnest on the similarly aged faunas from the Hell Creek Formation, as well as the Paleocene faunas in the overlying Tullock Formation in McCone County, eastern Montana. Several papers by Robert Sloan and Leigh Van Valen (Sloan and Van Valen 1965; Van Valen and Sloan 1965; Van Valen 1978) described mammals from both Cretaceous and Paleocene faunas in McCone County. Many of these mammals were from the Bug Creek area in McCone County. Sloan and Van Valen (1965) argued that Bug

Creek mammals were latest Cretaceous in age, but as described in chapter 3, the consensus is now that the Bug Creek faunas are Paleocene in age and include a considerable amount of reworked material. Richard Estes and colleagues investigated the lower vertebrates (meaning, nonmammals) and the paleoecology of the faunas described by Sloan and Van Valen (Estes et al. 1969; Estes and Berberian 1970). Largely because of their synthesizing work, a geographic pattern began to emerge for latest Cretaceous vertebrate faunas in the Western Interior. A decade later, working with Bill Clemens a short distance to the west in Garfield County, I examined the latest Cretaceous mammals (Archibald 1982). Howard Hutchison and I then examined the turtles (Hutchison and Archibald 1986). Finally, Laurie Bryant examined all nondinosaurian lower vertebrates (Bryant 1989).

Portions of approximately contemporaneous faunas were also described from farther north. Mammals were described by Jason Lillegraven in 1969 from the Scollard Formation of Alberta and by Richard Fox in 1989 from the Frenchman Formation of Saskatchewan. Fox dealt with all vertebrates from the Frenchmen Formations, but more expanded studies focusing on just the lower vertebrates (such as those conducted by Estes and Bryant to the south) have not yet been done on the Late Cretaceous Canadian faunas.

An important, but certainly not singular, goal of all these studies was to understand and explain what happened to the vertebrates at the K/T boundary. Leigh Van Valen and Robert Sloan were especially interested in this question; they published several papers (e.g., 1977) expounding upon the issue of extinction. Much of their analysis concentrated on the Bug Creek sequence, which they used to argue effectively for a gradual or stepwise turnover of vertebrates leading up to and crossing the K/T boundary. Most workers, including me, accepted their arguments until issues of reworking in the Bug Creek sequence (discussed in chapter 3) began to surface.

In part, suspicions of reworking came about when the catastrophic impact theory proposed by the Alvarez group in 1980 was seen to conflict with the then-current view of the vertebrate fossil record—that it showed a gradual pattern of extinction and replacement near the K/T boundary. The question of reworking is by no means settled today, but that issue has made it clear that careful reinterpretive work is exceedingly important.

Thus Laurie Bryant and I decided to reexamine the whole Hell Creek vertebrate fauna to see if any patterns could be detected that would better explicate events at the K/T boundary. We presented our results at the second Snowbird conference on impacts and extinctions, held in 1988 at the Snowbird Ski Lodge outside of Salt Lake City (Archibald and Bryant 1990). This conference was one of the most stimulating I have ever attended, in large measure because of its interdisciplinary nature.

As Bryant and I noted in our paper, the database for this study was more than 150,000 vertebrate specimens housed in the Museum of Paleontology at the University of California, Berkeley. The specimens came from localities that are unquestionably in either the uppermost Cretaceous Hell Creek or the lower Paleocene Tullock Formation in Garfield and McCone counties, eastern Montana. Thus there is little or no chance of substantive reworking.

We did not go into the collections at Berkeley and painstakingly examine and count all 150,000 specimens. Rather, we felt sufficient confidence could be achieved by thoroughly examining well-preserved specimens and estimating abundances by using smaller samples. These estimates were based on what we knew was the fossil yield from most localities. During the years of fieldwork we had kept accurate records of the amount of fossil material that had been collected. From this and our own earlier published work on the fauna (Archibald 1982; Bryant 1989) we were able to establish a rough—but still very useful—approximation of the number of individuals belonging to each species. These estimates were used only to consider the general issue of species rarity in our paper, "Differential Cretaceous-Tertiary extinctions of nonmarine vertebrates: Evidence from northeastern Montana."

Using the data from the Berkeley collections and reports of some other rare species in other museums, we recognized 111 species of vertebrates within the uppermost Cretaceous Hell Creek Formation in Garfield and McCone counties. As the word "differential" in the title of our 1990 paper indicated, we found that different groups of vertebrates were affected differently across the K/T boundary. This differential pattern of survival has a profound effect on assessing the various corollaries of the major extinction theories—a task I will attend to later. Our results showed that overall species survival for vertebrates across the K/T boundary in this region was between 53 and 64% (not the 32% mistakenly reported for our results by Ward 1994). The span of values reflects our attempt to accommodate three artifacts of the fossil record that could easily have skewed the data. These artifacts (discussed at length in chapter 4) are local extirpation, pseudoextinction, and rarity.

Bryant and I considered the issue of survival outside the study area. We realized that when a species survived the K/T boundary elsewhere in North America but disappeared in eastern Montana, the recorded extinction in our data set was only local extirpation. The species had obviously survived, although its geographic range was diminished. Thus we did not consider such local extirpations as true extinctions. To this extent, we used knowledge from the full geographic reach of the fossil record to enhance the reliability our own, smaller data set.

We also realized that when a species disappears from the fossil record through the process of speciation, either through cladogenesis (splitting) or anagenesis (change within a single lineage), its disappearance could not be taken as true extinction—that is, as the species being killed off or failing to reproduce for one reason or another. Rather, this is one form of pseudoextinction. If one is not careful to extract the pseudoextinctions from the true extinctions, then conclusions about "extinction" rates can be greatly exaggerated. A case in point: in my study of early Tertiary mammals (Archibald 1993a) I found that in about 25% of the cases, species disappearances were the result of speciation events—not extinction events. In our 1990 paper, Bryant and I referred to the issue of pseudoextinction as differences in rates of evolutionary turnover, simply meaning that some lineages will speciate more rapidly than others. If not considered, disappearances of more rapidly evolving lineages would have been erroneously treated as true extinction.

Bryant and I calculated an extinction figure of 96% across the K/T boundary for mammals as a group in our northeastern Montana data set. No detailed phylogenetic analyses are available for most latest Cretaceous mammals, but a modicum of common sense tells us that this 96% disappearance should not be taken as true extinction. Mammals were beginning a tremendous radiation that occurred after the disappearance of dinosaurs, and thus what would be treated as disappearances are pseudoextinctions resulting from speciation. Pseudoextinction is undoubtedly masking as true extinction in other vertebrate groups, too. Until we have better phylogenies for these other animals, the true extent of pseudoextinction remains hidden.

In a recent book on mass extinction, author Peter Ward (1994:158), failed to make this distinction between true extinction and pseudoextinction: "Contrary to popular opinion, however, mammals did poorly; in the Hell Creek region only one of twenty-eight species of mammals survived. In all North America, the survival rate of the mammals was only about 20 percent." In my view, such low numbers indicate a rather simplistic and too-literal reading of the fossil record. Ward, however, adopted a far more sophisticated approach in his argument that ammonite extinction at the K/T boundary was catastrophic. (Ward is, after all, an ammonite specialist.) In many sections, ammonites disappear well below the K/T boundary. Ward did not regard this disappearance as real, invoking the Signor-Lipps Effect. As you may recall from chapter 4, the Signor-Lipps Effect holds that the highest occurring specimen in a geological section probably is not the point of extinction. We are unlikely to find the remains of the very last individuals just before extinction simply because our sampling is not perfect. When one attempts to correct the fossil record for biases, it must be done evenhandedly, not simply to bolster one's own point of view.

Finally, Bryant and I struggled with the question of rarity of taxa. Although it was admittedly arbitrary, we placed the cutoff at fifty individuals. Any species represented by fewer than fifty specimens in the 150,000-specimen Berkeley sample would be deemed rare. It was over this issue that we pondered, discussed, and argued the most, even offering in our 1990 paper different views of how this issue should be handled. Rarity is thus the least tractable of the three artifacts that we believed must be taken into account when interpreting the fossil record.

Bryant and I found a tremendously higher percentage of disappearance for rare species. Of the rare species, 91% disappeared and presumably became extinct. Among the more common (or less rare), disappearance ran at 54%. Why the disparity? I then suggested it was a strong preservational bias and that we should exclude the rare taxa from the analysis (later in this chapter you will see that I have since changed my mind). My coauthor felt that the higher percentage of extinction for rare species was real. Rarer species might simply be more vulnerable to whatever factors were driving extinction. How to treat rare species in any such analysis remains a problem—exclusion ignores part of the fauna, while inclusion may overemphasize the importance of truly rare taxa. I have elsewhere offered a slightly different approach (Archibald 1995), which I discuss in the next section.

One approach to dealing with these artifacts of rarity and local extirpation was to count as survivals only common latest Cretaceous vertebrates both within and outside the study area (Archibald and Bryant 1990, table 2). This subset of the basic data resulted in a survival level of 53% (36 of 68 common species of vertebrates both within and outside the study area). When the pseudoextinction of both rare and common mammals was included, survival of vertebrates increased to almost 64% (47 of 74 species). These two estimates provided the bounds for the range of vertebrate survival that we calculated. Thus we concluded that between 53% and 64% of vertebrate species probably survived across the K/T boundary in eastern Montana.

Overall we concluded that although some patterns of extinction and survival could be recognized (e.g., aquatic versus terrestrial, large v. small), no single, dominant pattern emerged. In reanalyzing a culled portion of the original data set, Sheehan and Fastovsky (1990) found a 90% survival for freshwater species of vertebrates and only 12% for terrestrial species. In response, and using a complete data set, I calculated (1993b) that the survival of freshwater versus terrestrial species was 78% and 28%, respectively. Thus, the Sheehan and Fastovsky culled data set had resulted in a 30% exaggeration between freshwater and terrestrial survival. As discussed in chapter 8, this pattern of differential extinction turns out to be one of four that have so far been recognized.

THE PRESENT STUDY

In the years following the 1990 publication of the Archibald and Bryant database for vertebrates across the K/T boundary in northeastern Montana, I continued to add new species as they were described, update any new occurrences, and eliminate highly questionable species. Table 5.1 provides the most recent list of species for the uppermost Cretaceous Hell Creek Formation. This table is, however, simplified from what Bryant and I presented in 1990. It shows only whether a particular species or species lineage survived the K/T boundary, and whether it was rare to begin with.

The only major difference in approach between this paper and our 1990 paper is in the treatment of rare species. In our 1990 table we calculated survival rates both with and without the rare species. With a few exceptions discussed here, the species in the updated table do include those that are rare, and I do not cull the rare ones from this list when I tabulate species turnover. Quite frankly, the reason for including the majority of rare species in the study is that I find no compelling biological reason to exclude them. The other two artifacts that Bryant and I considered in analyzing turnover (survival of species outside the study area and survival of lineages through speciation) are, however, incorporated in table 5.1, just as they were in our 1990 table. For example, the multituberculate *Mesodma formosa* is depicted in table 5.1 as surviving the K/T boundary because specimens have been recorded from Paleocene sediments outside our study area. The eutherian *Gypsonictops illuminatus* is depicted in table 5.1 as surviving the K/T boundary because it is interpreted to be the nearest relative of early Tertiary species and thus did not disappear without issue.

Recall that in our 1990 paper Bryant and I maintained that 43 of our

111 species of vertebrates were rare (numbering fifty or fewer specimens in the full 150,000 sample). Of these 43 rare species, 28 seem to disappear from the record at the K/T boundary, while 15 species or species lineages survive in the study area or elsewhere. Basically following Bryant's lead, I now accept the view that the disappearance of most rare species probably signals true extinction. Note that I indicated "most," because I now identify 8 of the 28 disappearances of rare species as problematic (see the notes to table 5.1). The problematic species either were so poorly known or were based on such questionable taxonomy that they still should be excluded from the analysis. This left 20 of 28 rare species disappearances that I felt relatively confident about regarding as true extinctions. As Bryant had suggested, these species may have already been rare or in decline. Events at the K/T boundary simply finished them off.

Among the problematic rare species that I excluded from the analysis was *Ugrosaurus olsoni* (CoBabe and Fastovsky 1987). This new genus and species of ceratopsian dinosaur from the Hell Creek Formation was based only on a single, partial specimen—the nose horn section of the skull. In 1993, however, it was shown that this particular horn actually belonged to the well-known *Triceratops* (Forster 1993).

A more interesting example of preservational rarity is the baenid turtle *Thescelus insiliens*. In 1986 Howard Hutchison and I published our examination of the paleoecology of fossil turtles from the uppermost Hell Creek and the lowermost Paleocene Tullock formations in eastern Montana. Because of their shell and mostly aquatic habits, we had a good collection of skulls, shells, vertebrae, and limbs of these turtles. We were able to assign almost seventy-five specimens, each to one of six species of the extinct family Baenidae. Only one Hell Creek specimen, however, was assignable to *Thescelus insiliens* (Hutchison and Archibald 1986). *T. insiliens* may actually not have been such a rare turtle, but it was known only from distinctive shells. In order to identify this species, we needed the associated front ends of both the upper and lower parts of the shell. Because we required such a specific portion of the shell, it is no surprise that only one of seventy-five specimens could be identified as *T. insiliens*. I thus excluded this species from the tabulations (table 5.1). Another interesting aspect of this turtle is that it is the only species of baenid not known to survive the K/T boundary. Given what we know of the other members of the family, it would not be a surprise should a specimen be found in the lowermost Paleocene Tullock Formation.

Of the 107 vertebrate species or species lineages that I currently recognize from the Hell Creek Formation, 49% (52 of 107) survived across the K/T boundary in the Western Interior. I regard this figure as the minimum percentage survival of vertebrate species across the K/T boundary. This is because fully twenty rare species that disappear and are presumed extinct at the K/T boundary are counted among the fifty-two extinctions. But these are species represented by fewer than fifty identifiable specimens out of 150,000. Although I cannot provide any accurate estimate, certainly some of these very rare species did survive. The extreme and very improbable case would be if all twenty rare species survived. If such were the case, then the survival rate would be 67% (72 of 107). This provides the extreme maximum percentage survival in the region. If I were to hazard an educated guess (and it is no more than a guess), I would say that no more than 60% of vertebrate species present in the latest Cretaceous Hell Creek Formation survived the K/T boundary.

In the discussion to follow I use the minimum percentage species survival of 49%. By itself, however, this percentage tells us little. We need to be able to compare it to changes in vertebrate species numbers at times further removed from the K/T boundary. Unfortunately, we do not have comparable species-level studies for the Late Cretaceous before the time of the Hell Creek Formation and for the early Tertiary after the Tullock was deposited. Bryant and I reported in our original study in 1990, however, that at the next coarser taxonomic level, there was only about a 10% lower generic survival at the K/T boundary compared to sampling intervals slightly before and slightly after the K/T boundary (figure 6.1).

What we did was to look at generic-level survival of vertebrates from the mid-Campanian Judithian mammal age into the late Maastrichtian Lancian mammal age, then from this mammal age into the early Paleocene Puercan mammal age (the K/T boundary), and finally from this mammal age into the middle Paleocene Torrejonian mammal age. These generic-level data were gathered from a host of publications (Archibald and Bryant 1990) by experts on Late Cretaceous and early Tertiary vertebrates. The generic-level survival figures are most reliable for the K/T and for the transition from the early to mid Paleocene because these represent contiguous mammal ages with durations of less than five million years. The survival figures from the mid Campanian across to the late Maastrichtian are probably not completely comparable, as they span almost ten million years, including the poorly known "Edmontonian" mammal age.

Even with these caveats the level of extinction for vertebrates across the K/T boundary is not the mass annihilation as it is often por-

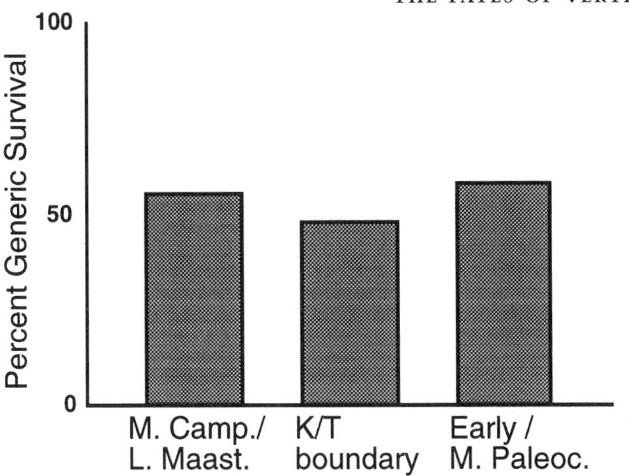

FIGURE 6.1 The K/T boundary in perspective. The percentage of genera of vertebrates surviving the K/T boundary in the western United States is only about 10% lower than the generic survival rate at sampling intervals before and slightly after the K/T boundary. *Source: Archibald and Bryant 1990.*

trayed. As I discussed at length in chapter 3 (pertaining to the studies of Sheehan and Fastovsky), using higher taxonomic groupings such as genera can lead to erroneous conclusions. Nevertheless, the little evidence that is available suggests that the presumed mass extinction at the K/T does not apply to vertebrates. What does occur at the K/T is the tremendous reorganization of the terrestrial component of the vertebrate fauna. With the apparent extinction of all nonavian dinosaurs (19 species were present in the Hell Creek, but none made it into the Tullock), mammals begin to dominate the larger-body vertebrate niches. This resetting of the evolutionary stage takes place in less than one million years, probably sooner.

The eventual replacement of dinosaurs by mammals is part of an even broader evolutionary and ecological pattern spanning the K/T boundary. The pattern of extinction and survival was not evenly distributed across the various vertebrate groups. The replacement of dinosaurs by mammals was one result of this uneven, or differential, pattern of extinction.

When plotted as in figure 6.2, the disparity in survival and extinction across the twelve major monophyletic vertebrate groups becomes obvious. Just five of the twelve groups—elasmobranchs, marsupials, squamates, ornithischians, and saurischians—together account for 75% of the extinction. What do sharks, lizards, and dinosaurs, and marsupials have in common that made each suffer 70% or more species extinction across the K/T boundary in eastern Montana? To even begin to understand the causes of the extinctions as a whole, we must explain this disparate pattern of extinctions. What better way to examine these observed patterns than to test them against the various corollaries (and proximate causes) of extinction, as would be predicted

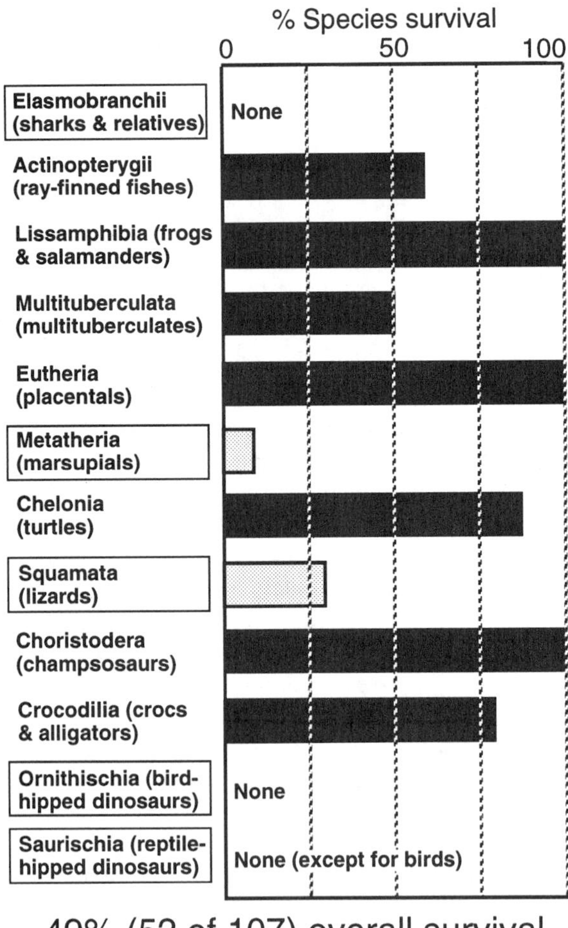

FIGURE 6.2 Survival patterns at the K/T boundary. Data for species of the twelve major monophyletic vertebrate groups included in this study are from northeastern Montana. Just five groups—elasmobranchs, marsupials, squamates, ornithischians, and saurischians—contributed 75% of the extinction.

from each of the three main extinction theories—that of impact, volcanism, and marine regression.

Table 6.1 builds on the calculations presented in figure 6.2 in order to compare the levels of extinction for the twelve major vertebrate groups against the corollaries of the three extinction theories, as well as some corollaries that are not theory specific. Note first that in figure 6.2 the twelve major taxa can be grouped into two categories—those that largely survived the end-Cretaceous extinction and those that largely did not. Only the multituberculates (with a 50% survival rate) cannot be so distinguished. There is no definitively biological level at which extinction goes from what one might consider minor to major, but the disparity among these twelve groups is obvious. In table 6.1, if species extinction was 70% or greater, it was considered a major extinction for that group. Five groups show 70% or greater extinction. The remaining seven groups have extinctions of 30% or less (50% for multituberculates).

I next asked, If a particular corollary, or proximate cause, of extinction did occur, what would be the predicted pattern of extinction for species within each of the twelve major vertebrate groups? As with the observed patterns, the predicted patterns of extinction resulting from the corollaries were assessed as to whether they are major or minor. The assessment was based on what is known of modern biotas, but there is no way we can provide estimated numbers or percent extinctions. We can, however, make some reasonable and qualitatively accurate general predictions as to how a group of vertebrates would respond if a given corollary were to occur, especially relative to other groups of vertebrates. This is in part made possible because, as we will see in the next two chapters, some of these corollaries are truly nasty events that would have caused very broad, devastating consequences, while others are quite narrow for groups that they might affect.

Table 6.1 Testing Possible Causes of K/T Extinctions Against Survivorship Predictions and Observed Patterns

	Sharks & relatives	Bony fish	Lissamphibians	Multituberculates	Placentals	Marsupials	Turtles	Champsosaurs	Lizards	Crocodilians	Bird-hipped dinos.	Reptile-hipped dinos.	Number of correct predictions
OBSERVATIONS:													
Number L K vert. species	5	15	8	10	6	11	17	1	10	5	10	9	
Number of K/T survivals	0	9	8	5	6	1	15	1	3	4	0	0	↓
Significant extinction	YES	NO	NO	NO	NO	YES	NO	NO	YES	NO	YES	YES	↓
ULTIMATE CAUSE:													
Impact (& volcanism?)	YES	yes	yes	yes	yes	YES	yes	yes	YES	yes	YES	YES	5
Proximate corollaries													
Sudden cooling	no	NO	yes	NO	NO	no	yes	yes	YES	yes	no	no	4
Acid Rain	YES	yes	yes	NO	NO	no	yes	yes	no	yes	no	no	3
Global wildfire	YES	yes	yes	yes	yes	YES	yes	yes	YES	yes	YES	YES	5
ULTIMATE CAUSE:													
Marine Regression	YES	NO	NO	NO	NO	YES	NO	NO	no	NO	YES	YES	11
Proximate corollaries													
Habitat fragmentation	no	NO	NO	NO	NO	no	NO	NO	no	NO	YES	YES	9
Lengthen streams	YES	NO	NO	NO	NO	no	NO	NO	no	NO	no	no	8
Competition	no	NO	NO	NO	NO	YES	NO	NO	no	NO	no	no	8
Corollaries not ultimate-cause specific													
Local wildfire	no	NO	NO	yes	yes	YES	NO	NO	YES	NO	YES	YES	9
Detrital influx	no	NO	NO	yes	yes	YES	NO	NO	YES	NO	YES	YES	9

Note: First row shows number of vertebrate species from the latest Cretaceous (LK) of eastern Montana. Second row shows how many of the LK species (and species lineages) survived the K/T boundary. Third row shows which major monophyletic taxa experienced significant species extinctions (<30% survival = YES; >50% survival = NO). The next eight rows predict whether each major monophyletic taxon should (YES, yes) or should not (NO, no) experience sigificant species extinctions as a result of each proximate corollary and as an aggregate of associated corollaries caused by two (three?) ultimate causes of mass extinction impact (and volcanism?) and marine regression. The last two rows are not specific to any ultimate cause. Capitilized YES and NO signify agreement between predictions of theory and observed patterns, while lower case yes and no signify lack of agreement. Numbers in right column indicate for how many of the twelve major monophyletic taxa predictions for levels of extinction agree with the observed.

VII

Shades of Dante's *Inferno*

The K/T impact turned the Earth's surface into a living hell, a dark, burning, sulfurous world where all the rules governing survival of the fittest changed in minutes.
Alan Hildebrand

The idea of an extraterrestrial impact causing extinction at the K/T boundary and possibly at other times during Earth history has become the darling of the media. It is a simple event. It evokes terror. It can happen again at any time. These are the bits of ideas that do well as sound bites and make spectacular reconstructions for the news. Such simplifications, however, do not do justice to the wealth of data that scientists have accumulated, both pro and con, about impacts. But the media are not alone in their oversimplifications and sensationalism. The introductory quote is not from a tabloid TV show; it is the main conclusion of a review article by a well-respected impact proponent, Alan Hildebrand (1993:112). Thus the task of sorting hoopla from evidence about impacts (and volcanic eruptions) and possible effects becomes all the more difficult.

DEATH FROM ABOVE: THE IMPACT THEORY

Present research on a possible K/T impact dates only from 1980, with the seminal paper in *Science* by the late Nobel laureate physicist Luis Alvarez, his geologist son Walter, and two nuclear chemist colleagues, Frank Asaro and Helen Michel. The argument that impacts might cause extinction, however, dates to at least the mid-eighteenth century, according to Hildebrand (1993).

Various kinds of physical evidence point to the occurrence of an impact. The first to be recognized is the enrichment at the K/T boundary of platinum-group elements, in particular, iridium. As Earth formed, heavier elements including those in the platinum group concentrated in the mantle and core and are thus depleted in the more surficial rocks of the crust. Higher percentages of such elements are found in volcanic rocks that tapped sources deep in the Earth and in extraterrestrial sources that retained the proportions found in the early Earth. Because of the very high levels of iridium at the K/T boundary compared to background levels (figure 7.1), the Alvarez group suggested the iridium enrichment was from an impact of an asteroid about six miles in diameter. Iridium anomalies were first recognized in surface exposures of marine rocks preserving the K/T boundary. Such anomalies were later also reported from drill cores taken through the K/T boundary in the deep ocean and from surface exposures of continental rocks. The latter included sections in rocks in eastern Montana yielding the vertebrate fossils I am here recounting. By 1990 Walter Alvarez and Frank Asaro could report that ninety-five sites from around the world had yielded an iridium anomaly at the K/T boundary.

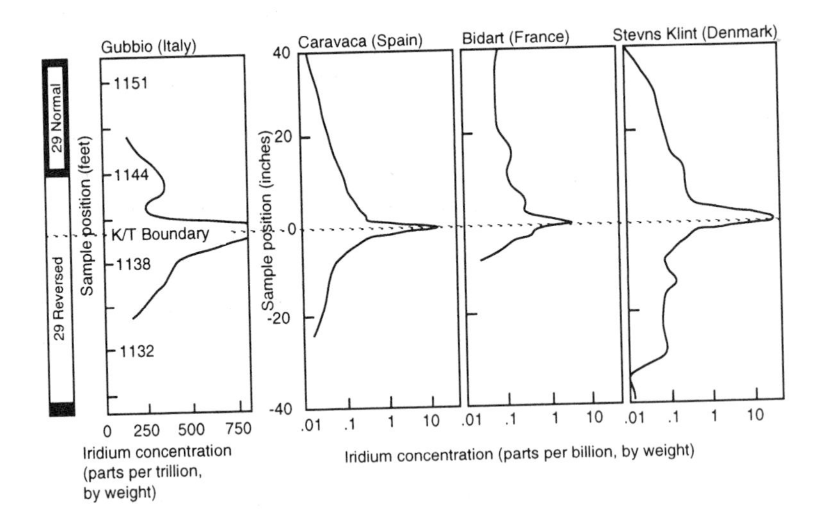

FIGURE 7.1 The iridium spike at the K/T boundary. Enhanced concentrations of the element iridium at the K/T boundary, compared to background levels for various sites around the world. The portion of the magnetostratigraphic column bracketing the K/T boundary is on the left. After Courtillot 1990.

Although iridium enrichment can be caused by volcanic activity, other physical phenomena were later reported that argued for impact. Probably of most interest were spherules and shocked quartz. The spherules were first reported in marine rocks in Spain in 1981 (Smit and Klaver 1981), but have been found elsewhere around the world. Although some of the spherules turned out to be fossil algae or even recent insect eggs (Courtillot 1990), others appear to be solidified droplets of molten rock possibly injected into the atmosphere by an impact.

Shocked quartz was first reported at the K/T boundary by Bruce Bohor and colleagues (1984). The so-called shocking is thought to occur when quartz is subjected to very rapid pressure and temperature increases. At lower pressures deformation of the quartz causes planes of shocking, or shock lamellae, aligned in one direction. As deformation increases, so too do the number of different planes of shocking (figure 7.2). The Bohor group attributed the shocked minerals that they examined to an impact. After the first reports, it was argued by some (e.g., Carter et al. 1986) that these multiple sets of planar deformation in quartz or even other minerals could have been caused by explosive volcanism. The issue has not been resolved to everyone's satisfaction, but because of the global distribution of these minerals showing multiple sets of planar deformation at the K/T boundary, a very large impact is the odds-on favorite.

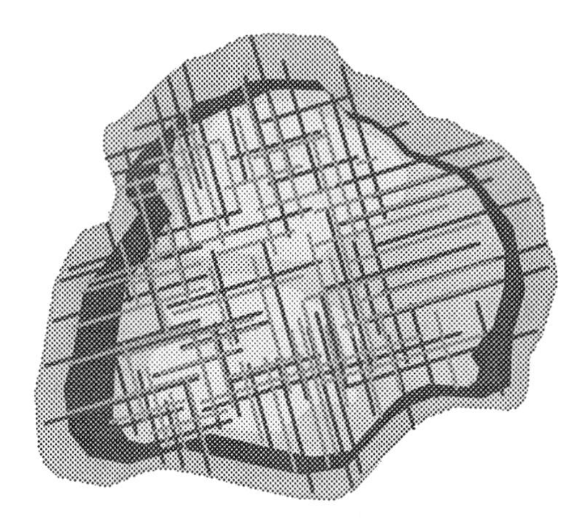

FIGURE 7.2 Shocked quartz. Sketch of the cross-section of a quartz grain with multiple sets of planar deformation caused by very great pressure from a substantial explosion (impact or perhaps volcanic).

From the first reports by the Alvarez group in 1980, various techniques were used to attempt to locate a crater at the K/T boundary. A number of candidates came and went, but a possible crater at the northern tip of the Yucatán Peninsula in southern Mexico seemed to fit most criteria (figure 7.3). The final piece of evidence that seemed to clench this site as ground zero was a radiometric date made on rock from the crater that had melted during the presumed impact, as well as ejected rock from nearby. This date of 64.98 (± 0.05 million years) was reported by Carl Swisher and colleagues (1992), and it was very close to the widely accepted date of around 65 mya for the K/T boundary. A number of earth scientists were involved in the detective work tracking this crater. In his fascinating account of this work, Alan Hildebrand (1993) reports that he named the crater Chicxulub after a nearby town and because this Mayan word means "tail of the devil" or "sign of the horn."

FIGURE 7.3 Site of a K/T impact? The Chicxulub structure at the northern tip of the Yucatán Peninsula in southern Mexico has been identified as an impact crater and dated near or at the K/T boundary.

Other physical evidence reported in support of an impact are nearshore marine deposits interpreted by some as caused by very large impact-generated waves. These also have been called tsunami deposits, after the giant waves sometimes generated by earthquakes and which, in the extreme, wash away whole seacoast towns. Tsunami deposits more than ten feet thick have been reported around the Gulf of Mexico. They are said to be unique to the K/T boundary. Although some of the beds may have formed by storms, earthquakes, or even impacts, caution must be used before any catastrophic interpretation is accepted. If they were suddenly formed, there should be no signs of burrowing sea creatures within the beds. In an open letter (1994), Tony Ekdale, an expert on marine ichnofossils (traces, tracks, and burrows), wrote that burrows do occur within some of these supposedly single-event tsunami beds. Ekdale found that the tops of some burrows were eroded before the next sediment was deposited. This means that these beds could not have been deposited by a single catastrophic tsunami wave; rather they represent multiple deposition events. Ekdale also indicated that these beds at the K/T boundary are not unique. Very similar sandstone units are found well below the K/T boundary in eastern Mexico. Similar K/T boundary sands in Alabama have been interpreted by Charles Savrda (1993) as the result of multiple events. Savrda's studies led him to "refute the tsunami origin" of these beds in Alabama.

Turning to biological ramifications, the original paper by the Alvarez group in 1980 still offers the basic mechanism of how such an impact might cause extinction among terrestrial species, including vertebrates. The impact would create a dust cloud enveloping the globe and that would last from a few months to a year. The world would be left in darkness for whatever length of time the dust remained in the atmosphere. Without sunlight, photosynthesis in the sea and on land would cease. As the plants died or at least became dormant, herbivores would starve, with carnivores not far behind.

Although a number of corollaries to this original scenario have been proposed to account for the mass extinction at the end of the Cretaceous, a common but incorrect perception is that all major taxa show very high levels of extinction across the ecological spectrum. Species extinction estimates of 75% (Glen 1990) or more are suggested, but I have yet to see any such estimates derived from species-level data; all such estimates are extensions of data for higher-level taxa. As I have emphasized, such extensions are suspect. The impact-generated scenario of very high levels of extinction across most environments is so broad-spectrum and tries to explain so much that it is difficult to test. In his excellent essay on catastrophic versus noncatastrophic theories of dinosaur extinction, Mike Williams (1994) notes that the burden of proof for such sweeping, catastrophic extinction scenarios should rest with the proposers of the theory. He argues that the proposers have thus far failed to shoulder this burden of proof in the case of dinosaur extinction.

We cannot, of course, reject outright the possibility that events such as the impact of an asteroid or comet at the K/T boundary caused all extinctions recorded in the rocks at this time. In order to seriously entertain such a scenario, specific, testable hypotheses of causes of extinction must be provided. Walter Alvarez (1986:653) suggested that "a whole Dante's Inferno of appalling environmental disturbances" accompanied the impact. Some of the major corollaries are a short and sharp decrease in temperature, highly acidic rains, and global wildfire. These are the corollaries that should be tested against the pattern of vertebrate turnover, as I have presented in table 6.1. I did not include in that exercise the central effect of the original impact theory, an extended interval of darkness or dusk. This is because, unlike the three corollaries just listed, it is difficult to estimate the direct effect of reduced light on vertebrates, other than possibly impairing vision for some. Most would agree that this central event would have had a more indirect effect on vertebrates, namely prompting the loss of primary production of plants. In the final chapter I will bring in other species, including plants, and discuss their pattern of extinction and its influence on the vertebrates.

A Sudden Frost

Although a short, sharp decrease in temperature was not emphasized in the original paper by the Alvarez group in 1980, it soon became an important corollary. As Walter Alvarez (1986) noted, work by atmospheric scientist Brian Toon and colleagues (1982) suggested that tremendous amounts of dust would be injected into the atmosphere after a large impact. "The darkness would also produce extremely cold temperatures, a condition termed impact winter" (Alvarez and Asaro 1990:81). According to Walter Alvarez (1986; Alvarez and Asaro 1990) this work on impact winter was a contributing factor leading to later research by Toon and colleagues on nuclear winter (Turco et al. 1983). Brian Toon and colleagues (1982) calculated that following a large impact, ocean temperatures would decrease only a few degrees because of the huge heat capacity of the oceans. On the continents, however, temperatures would be subfreezing everywhere for at least forty-five days and perhaps as long as six months. In general, the temperature would remain subfreezing for about twice the time of darkness caused by the dust.

Does the fossil record support the argument for a prolonged, severe cold snap? According to the paleobotanists Jack Wolfe and Garland Upchurch (1986), the answer is yes. They claim that biological data are consistent with a period of from one to two months of a mean temperature near freezing (0°C). Wolfe (1991) has gone even further, claiming that this "impact winter" occurred "approximately in early June." At least the latter argument has met with very little or no support among other paleobotanists (Nichols et al. 1992). This claim may be true for the fossil plant record, but it is not substantiated in the case for vertebrates—quite the opposite.

If such a prolonged regime of summertime subfreezing did occur, which vertebrates would have been affected? If we look to the living counterparts of those that lived at the end of the Cretaceous, we see that, in general, ectothermic tetrapods would likely suffer most. The reasoning demands a bit of explanation.

Ectotherms, as the name suggests, heat or cool themselves using the environment. Endotherms such as mammals and birds generate their heat through metabolic activity. In endotherms somewhere around 80% of food consumption goes toward thermoregulation (regulation of body temperature). Fish, which are by and large ectothermic and whose aqueous habitat dulls the immediate severity of atmospheric temperature excursions, would be generally less affected by a severe temperature drop.

Howard Hutchison (1982) noted that the northernmost reaches of extant turtles and crocodilians is limited by temperature. None can follow the mammal and bird clades into high latitudes. Further, turtles and crocodilians cannot tolerate freezing, and they become sluggish or immobile at 10–15°C. Today, some lissamphibians (frogs and salamanders) and reptiles inhabiting areas with low winter temperatures or of severe drought have evolved methods of torpor (estivation and hibernation) to survive inclement times. These are the exceptions, however, because species diversity for ectothermic tetrapods is far higher in warmer climates. Further, there is no basis to assume that Late Cretaceous ectothermic tetrapods living in subtropical to tropical climates—such as existed then in eastern Montana—were capable of extended torpor. Moreover, torpor is most often preceded by gradual decreases in ambient temperature, changes in light regimes, and decreases in food supply. The ectotherms in eastern Montana could not have anticipated a short, sharp decrease in temperature.

This is true even if the impact had occurred during a Northern Hemisphere winter, as winter temperatures would normally have been only slightly lower than during the summer months (the Cretaceous is widely believed to have been a time of greenhouselike balminess, at least compared to our modern climate). We must remember that this was a subtropical to tropical setting—despite the latitude—and thus the extended subfreezing temperatures advocated by proponents of the "sudden frost" corollary of the impact theory would have been devastating to ectotherms even during a Late Cretaceous winter in Montana.

Figure 6.2 and table 6.1 showed that except for a 70% decline in lizards, ectothermic tetrapods (frogs, salamanders, turtles, champsosaurs, crocodilians) did very well across the K/T boundary. The corollary of a short, sharp drop in temperature thus does not accord well with the vertebrate data at the K/T boundary. Far more of the ectotherms should have been affected. As table 6.1 shows, only four of twelve major vertebrate taxa have extinction and survival patterns that fit this thesis of the biological effects of such a temperature decrease. The problems with the contention that a severe cold spell marked the K/T boundary go beyond its lack of support from the vertebrate fossil record. As David Raup (1991) notes, we do not even know whether an atmosphere filled with dust would, as this corollary suggests, initiate a severe cooling because of reduced sunlight, or a greenhouse effect because of trapped heat.

The argument for a short, sharp temperature decrease at the K/T boundary became even shakier in 1993 when Bill Clemens and Gayle Nelms described a latest (but not terminal) Cretaceous vertebrate fauna. This fauna came from along the Colville River in Alaska at about latitude 70°N, which was probably at latitude 85°N during the Late Cretaceous—in either case, well above the Arctic Circle. Thus, any organisms there during the northern winter (which included the plants, which were certainly not going anywhere) faced more than three months of dusk or darkness as a normal annual event.

Clemens and Nelm compared their Alaska fauna with the compilation that Bryant and I (1990) had made for eastern Montana. Figure 7.4 illustrates the results of their comparison. They were able to make comparisons only among genera, but even at this level a difference between the two faunas was obvious. They noted an aggregate of forty-nine genera of amphibians and nondinosaurian reptiles in the Montana group, but found none in their Alaskan fauna. Overall, the Alaska group is notably less diverse, with only fourteen genera compared to ninety-nine for eastern Montana, but the total lack of the two groups that compose almost half of the fauna from eastern Montana had to be more than an artifact. The simplest, and I believe correct, answer forwarded by these authors was that endothermic tetrapods (dinosaurs and mammals) and ectothermic fishes could deal with the lower temperatures in latest Cretaceous Alaska, while ectothermic tetrapods could not. As discussed in chapter 5, the question of dinosaur endothermy remains unresolved; but simply by virtue of larger size, most dinosaurs would have been able to maintain a higher than ambient temperature.

Further south, in the balmy climes of latest Cretaceous eastern Montana, ectothermic tetrapods were abundantly present. If an annual temperature range of 2–8°C (Clemens and Nelms 1993) was enough to exclude ectothermic tetrapods during the latest Cretaceous in Alaska, a severe bout of subfreezing temperatures at the K/T bound-

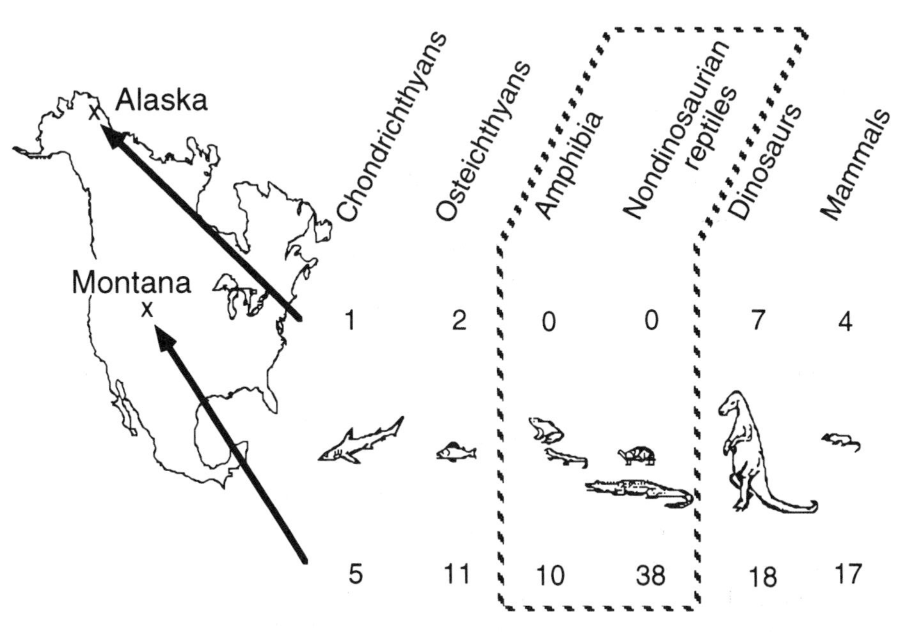

FIGURE 7.4 Latitudinal differences in Late Cretaceous vertebrates. Numbers of genera of the latest (but not terminal) Cretaceous vertebrate fauna from the Colville River in Alaska, compared to the terminal Cretaceous vertebrate fauna from the Hell Creek region of Montana. Colville River fauna is now at about latitude 70°N (probably 85°N during the Late Cretaceous), which is well above the Arctic Circle. Note the huge disparity in amphibians and nondinosaurian reptiles between the two faunas. After Clemens and Nelms 1993.

ary should have devastated the rich ectothermic tetrapod faunas at mid latitudes. As figure 6.2 shows, however, the opposite is true. These species flourished.

The Acid Test

As with other corollaries of the impact theory, the conjectured likelihood and importance of acid rain varies from author to author. The different ideas pertain to the kinds of acid that might be produced and the level of acidity.

Among the acids most commonly cited as likely products of an impact are nitric and sulfuric acid. Nitric acid (HNO_3) would be produced because the tremendous energy released by an impact allows the combination of atmospheric nitrogen and oxygen (Lewis et al. 1982). Sulfuric acid (H_2SO_4) would be produced as large amounts of sulfur dioxide are vaporized from rock at the impact site—especially if the impactor happened to land in a gypsum-rich area. (D'Hondt et al. 1994a). According to the impact scenario, both acids would be precipitated in the form of rain. Estimates of the pH of such acid rains vary. Prinn and Fegley (1987) suggested that, depending upon the impacting object, the global rains could have a pH as low as 0–1.5. D'Hondt et al. (1994a:30) suggested global effects could have "driven the pH of near-surface marine and fresh water below 3." The latter authors do seem to be aware that the patterns of vertebrate extinction do not fit their scenario, yet if I correctly understand them, they still say that the acid rains would have been "rapid" and "nonuniform."

For those of us who may have forgotten more chemistry than we have retained, a refresher of the concept of pH is in order. I looked to a chemistry colleague at my own university for guidance, Dewitt Coffey. Through him I learned to appreciate the significance of the truly staggering pH values presented in the last paragraph.

For most of us nonchemists, if we think about the pH scale at all, we remember it as running from the most acidic of 0 to the most basic of 14. Dewitt pointed out to me, however, that because it is a logarithmic scale, one can have values less than zero. The pH value refers to the concentration of hydrogen ions (H^+). Each unitary drop in the pH represents a ten-fold increase in the concentration of hydrogen ions. Pure water, with its pH of 7.0, is defined as neutral—neither acidic nor alkalinic. Water in equilibrium with today's atmosphere is slightly acidic, with a pH of 5.6. The oceans, however, are slightly alkaline, with a pH of around 8, owing to geochemical processes.

Another colleague at my university, the terrestrial ecologist George

Cox, recently completed a book on conservation ecology (Cox 1993), which includes a discussion of acid rain. He notes that when rain is below a pH of 5.0, it is considered unnaturally acidic. Rain as low as 2.4 has been recorded, but annual averages in areas affected by acid rain range from 3.8 to 4.4. Acid fogs and clouds from 2.1 to 2.2 have been recorded in southern California and have been known to bathe spruce-fir forests in North Carolina. Although different vertebrates react differently to chemical perturbations, aquatic species (fish, lissamphibians, and some reptiles) are the first and most drastically affected by acid rain. Fully aquatic eggs and young are the first to suffer, but if pH becomes low enough (e.g., lower than about 3.0), deaths of adults may occur. The affects on aquatic vertebrates across the K/T boundary would have been very bad if a pH of 3 was reached and truly horrendous if it hit 0 as suggested by some of the authors noted above.

One strong advocate of K/T acid rain is Greg Retallack, a scientist studying ancient soils (e.g., 1993). When I and others have pointed out to him that aquatic vertebrates, who would be most susceptible to the ravages of acid rain, have among the best levels of K/T survival, he has countered that the surrounding soils or bedrock would have buffered the aquatic systems. He even suggested (1994:1392) that "in such an acid rain crisis, limestone caves could have been important refugia for birds, mammals, amphibians, and small reptiles." The effects of acid rain can be buffered to some degree if the bedrock or soils are high in carbonates such as limestone. The problem with the refugia thesis is that there was no underlying carbonate bedrock in the latest Cretaceous of eastern Montana (or today, for that matter) that could have buffered the aquatic system. Further, because there was no carbonate bedrock underlying the aquatic systems, there could not have been any limestone caves in which creatures could have hidden. Carbonate cements do commonly bind the grains of sand within the uppermost Cretaceous sandstones, but these carbonates are clearly derived from groundwaters percolating through the sediments long after the K/T boundary. Even when bedrock is dominated by carbonate rocks such as limestone, this does not insure against highly acidic water. Such is the case today for many lakes in Florida (Pollman and Canfield 1991).

If pH became low enough to kill large nonaquatic vertebrates such as dinosaurs by inhalation or contact with skin or food, any possible buffering in aquatic systems would have been swamped by very low pH rain. When I used the analogy of battery acid for the pH of rain near 0, Dewitt Coffey responded (written comm. 1994) with a chemist's matter-of-factness, "Battery acid is actually about five times more concentrated than pH = 0 and about 160 times more concentrated than pH = 1.5. A pH of 1.5 is quite acidic, of course, and I would guess that essentially all life forms would die from significant exposure to pH = 1.5."

Given what we know of the modern biota's reaction to acid rain, aquatic animals would surely have been first and most to suffer. Yet there is almost no correlation between these predicted effects for aquatic vertebrate species at the K/T boundary and what the fossils say (Weil 1994a,b). Only three of the twelve major vertebrate taxa have extinction/survival patterns matching the acid rain corollary of impact (table 6.1), and thus the likelihood of low pH rain having been a significant cause of the extinctions is highly implausible. Even those studying the chemical basis for believing there was significant acid rain production following an impact are now questioning the acid rain corollary of impact because it cannot be reconciled with the fossil record (D'Hondt et al. 1994b).

More than Mrs. O'Leary's Cow

Another corollary of the impact theory that receives various levels of support is global wildfire. As with Mrs. O'Leary's cow which, as legend tells us, kicked over the lantern, starting a small blaze that grew into the great Chicago fire of 1871, something might have triggered a global conflagration.

Wendy Wolbach and colleagues (e.g., 1990) reported soot and charcoal from several sites at the K/T boundary coincident with the enrichment of iridium. If you recall, an iridium anomaly is one of the signatures of impact (and sometimes volcanism). The authors argued that this pattern was unique and must come from the extremely rapid burning of vegetation—equivalent to half of all the modern forests, according to Alvarez and Asaro 1990. A slightly different twist given by Ivany and Salawitch (1993) suggests that the isotopic signal of the carbon found in K/T boundary sections would require the burning of 25% of the above-ground biomass at the end of the Cretaceous. These authors, like the Wolbach group, stressed that wildfires at the terminal Cretaceous must have been a global phenomenon. Such a conflagration is really beyond our comprehension. Simply to grasp the magnitude of this corollary to impact, imagine a quarter to half of all structures on the globe engulfed in flames within a matter of days or weeks.

In such an apocalyptic global wildfire, much of the above-ground biomass all over the world, both plants and animals, would have been reduced to ashes. In fresh water, those plants and animals not boiled outright would have faced a rain of organic and inorganic matter unparalleled in human experience. These organisms would have choked on the debris or suffocated a short time later as oxygen was depleted by bacteria decomposing the tremendous influx of organic matter. The global wildfire scenario is so broad in its killing effects that it could not have been selective. Yet the fossils tell us that the vertebrate pattern of extinction and survival was indeed highly selective. It thus is no surprise that the global wildfire scenario only comports with five of the twelve cases of extinction/survival in table 6.1, and even this level of agreement occurs only because arguably all groups would have been equal opportunity losers.

The problem with the global conflagration corollary is not only its disagreement with the kind of biological selectivity (winners and losers) that one would expect. There is also a problem with the presumed physical basis for the event even occurring. Proponents argue that a global charcoal and soot layer coincides with the K/T boundary, whose emplacement is measured in months (Wolbach et al. 1990). Thus it also must be assumed that the sedimentary layer encasing the charcoal and soot was also deposited in only months. This is demonstrably not the case for at least one K/T section that continues to be cited in these studies—the Fish Clay of the Stevns Klint section on the coast of Denmark. As Officer and Ekdale (1986:263) explained, the Fish Clay is a laterally discontinuous, complexly layered and burrowed clay reflecting "the environmental conditions at the time of its deposition." It is not the result of less than a year of deposition caused by an impact-induced global wildfire. Thus, carbon near the K/T boundary at Stevns Klint as well as in other sections is likely the result of much longer term accumulation during normal sedimentation.

Summary

When all three corollaries of an asteroid or comet impact—short-term temperature decrease, acid rain, global wildfire—are summed as shown in table 6.1, the agreement of observed versus expected extinction/survival patterns is only five of the twelve major clades of vertebrates. Thus, the actual pattern of extinction and survival for vertebrates at the K/T boundary in eastern Montana is in relatively poor agreement with the corollaries of the impact theory. Unless a fatal flaw can be found in the assumptions I used to determine relative vulnerabilities to environmental perturbations, then these corollaries as currently proposed must be judged as unlikely causes of vertebrate extinction. This does not mean all corollaries of impact should be rejected even if some lose favor, but it is imperative that proponents of the different corollaries distinguish those that are supported by the vertebrate fossil record from those that are not.

DEATH FROM BELOW: THE VOLCANISM THEORY

As the impact and volcanism theories began to crystallize in the writings of their advocates, it became clear that both sides envisioned many of the same corollary events occurring at the K/T boundary. Companion pieces in the October 1990 issue of *Scientific American* entitled "What Caused the Mass Extinction?" show this symmetry well. Walter Alvarez and Frank Asaro argued the case for an extraterrestrial impact, and Vincent Courtillot argued it for volcanic eruptions.

As Courtillot notes (p. 89), "The appalling consequences of an asteroid impact and a massive volcanism would be quite similar." Some proponents of impact apparently dispute this view (e.g., Alvarez et al. 1994). Nevertheless, champions of volcanism such as Courtillot (at least in this article) argue that most of the physical events attributed to an impact also would occur during massive volcanism. The biological results would thus be similar. I find this argument highly plausible and therefore have aggregated impact and volcanism in the examination of ultimate and proximate causes depicted in table 6.2. Because the three corollaries of impacts just discussed would also pertain to volcanism, and because the predicted biological effects of these corollaries do not fare well when tested against the vertebrate fossil record, the volcanism theory, in my view, is equally suspect.

The volcanism theory leads to predictions that do differ in some ways from those that flow from the impact theory. Key differences pertain to timing. The impact theory calls for most of the major, cataclysmic effects to occur in just months or years, with residual physical effects possibly lingering for a few hundred or a few thousand years. For most volcanic eruptions, effects beyond the immediate area usually linger for a few months or a few years, such as after the eruption of Mt. St. Helens in 1980. There have been, however, episodes of very prolonged eruptions called *flood basalts*, the best known in the United States (and one of the more recent) being the sixteen million-year-old flows in the Columbia Basin of the Pacific Northwest.

In the past 250 million years, arguably one of the biggest flood basalt eruptions occurred on the Indian subcontinent. This was coincident with (and probably related to) the collision of the subcontinent with Asia. The most obvious manifestation of the collision that is still visible today is the world's tallest mountain range, the Himalayas. To the south, the flood basalts, known as the Deccan Traps, cover an immense part of both India and Pakistan (figure 7.5). Fifteen years ago I had the opportunity to visit a small but nevertheless impressive outlier of the Deccan Traps in Pakistan.

Vincent Courtillot (1990) notes the etymology for the term *Deccan Traps* and provides some sense of the huge scale of these flood basalts. *Deccan* derives from the Sanskrit word for "southern," and *trap* means "staircase" in Dutch. Although he does not say, I assume the allusion to a staircase comes from the giant stair-step effect created by the characteristic weathering pattern of these massive basalts. And massive they are. Courtillot notes that the most extensive individual flows in the sequence may cover almost 4,000 square miles, with volumes exceeding 2,400 cubic miles. Individual flows are generally 30 to 160 ft thick, sometimes reaching 500 ft. He notes that in western India the accumulations of lava flows may reach 7,800 ft, or a mile and a half thick. He suggests that the flows originally may have covered almost 800,000 square miles, with a volume possibly exceeding 350,000 cubic miles. For those whose eyes glaze at high numbers, consider this:the Deccan Traps contain almost enough basalt to cover both Alaska and Texas to a depth of 2,000 ft.

All this lava did not, however, spread out upon the earth all at once in a single horrendous catastrophe. A group of Indian and French geologists suggested in 1994 (Prasad et al. 1994) that the age estimates for

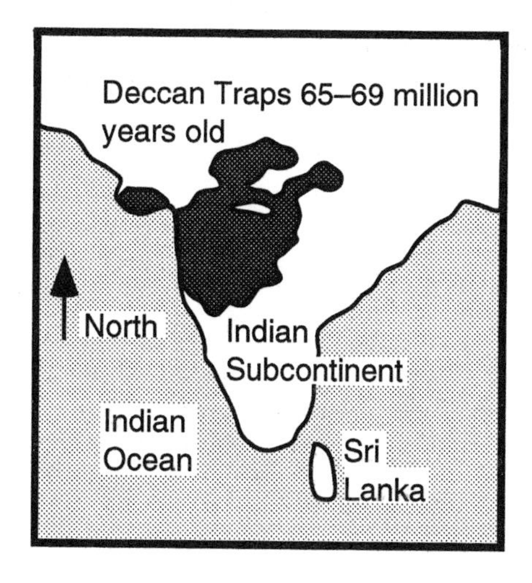

FIGURE 7.5 Flood basalts that straddle the K/T boundary. The Deccan Traps cover immense areas in both India and Pakistan. After Courtillot 1990.

emplacement of the Deccan Traps runs from about 69 to 65 mya. The eruptions, moreover, were episodic—not continuous. Courtillot suggests a peak eruption surrounding the K/T boundary. It is not clear precisely how close to the K/T boundary individual episodes can be placed. What can be claimed with considerable assurance is that the eruptions do bracket the boundary (figure 7.6). Based on radiometric dating, paleomagnetics, and vertebrate fossils, the bulk of the eruptions are believed to cluster around the K/T boundary, during a reversal in Earth's magnetic poles known as 29R or 29 reversed (Courtillot 1990; Prasad et al. 1994). The number 29 represents the 29th reversal of the Earth's magnetic field counting back from the present, which today has normal polarity by definition. The K/T boundary happens to fall in 29R. Changes in the Earth's magnetic poles have been examined as a possible cause of extinction, but no clear correlation was found. Charting the changes in magnetic polarity, or magnetostratigraphy, has been an invaluable tool in earth sciences, even though the phenomenon did not turn out to be a killer.

Although one cannot eliminate the possibility of sudden, violent eruptions, most of the Deccan Traps erupted more like the Hawaiian volcanoes rather than Mt. St. Helens, although one must never forget that the emplacement of the Deccan Traps dwarfs either of these. What would have been the effects on the global biota? One of the greatest effects would have been to increase and maintain a much higher level of particulate matter in the atmosphere. Whether those particulates would have caused warming through a greenhouse effect, cooling because of reduced light, or simply prettier sunsets is not certain. The additional carbon dioxide pumped into the atmosphere by the eruptions may have been a boon for green plants, which depend on this gas for photosynthesis. Increased carbon dioxide would also have contributed to the greenhouse effect. So, would the net effect have been a cooling or a warming? Would plants have been happier or no?

Oxygen isotope data suggest that at least regionally, if not the globally, climate cooled through the K/T transition (e.g., Lécuyer et al. 1993 estimate a decrease of 8°C in ocean temperature along northern Africa). Because the time frame is moderately long, many species on land and in the sea were probably able to adapt to the cooling. Those with short generation times likely would have had an advantage in evolving necessary adaptations, while taxa such as the dinosaurs may not have been as fortunate. Although the cooling recorded in the rock would have been an unlikely cause of extinction for most species at the end of the Cretaceous, it would have been an added stress.

A final and lasting effect suggested for eruption of the Deccan Traps

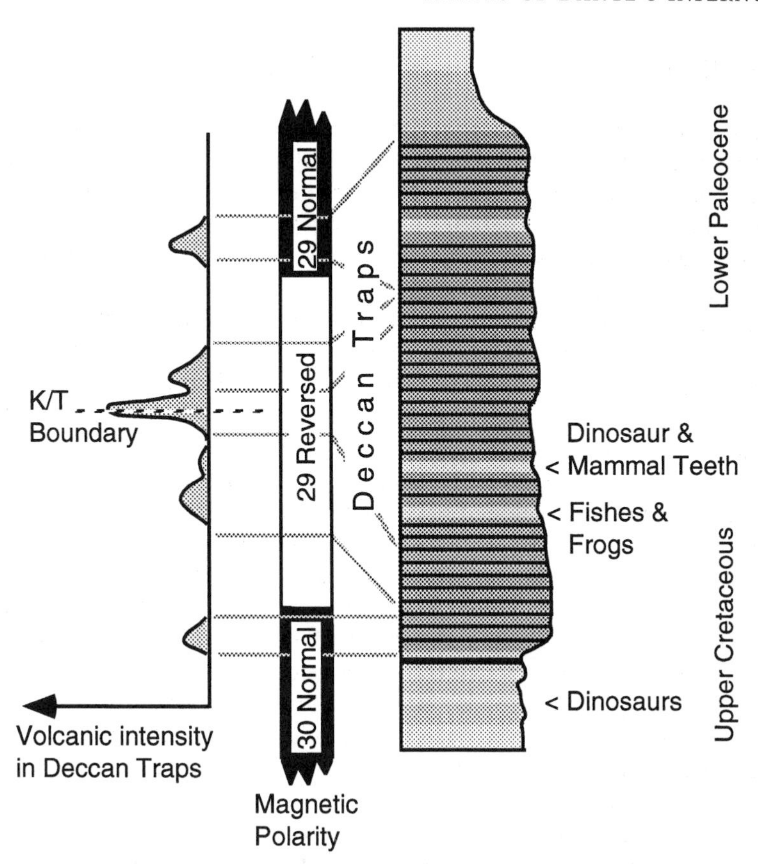

FIGURE 7.6 Correlating the Deccan Traps with the K/T boundary. The correlation chart indicates episodic eruptions of flood basalts that built the Deccan Traps 69–65 mya, with the peak eruption bracketing the K/T boundary. Based on radiometric dating, magnetostratigraphy (middle), and vertebrate fossils (right), it appears that the bulk of the eruptions are centered around the K/T boundary, during a reversal in Earth's magnetic poles known as 29R (29 reversed). After Courtillot 1990.

is reduced hatching success for eggs of herbivorous dinosaurs. Volcanic activity can release elements such as selenium that are highly toxic to developing embryos. Hans Hansen (1991) found increased levels of selenium in the eggshells of dinosaurs nearest the K/T boundary in southern France. Poisoning of eggs has also been reported from dinosaur eggs near the K/T boundary in Nanxiong Basin, southeastern China (Stets et al. 1995); various trace elements were found in the eggshells, but selenium was not one of them.

TAKING SERIOUSLY WHAT THE FOSSILS SAY

As should be very clear from the preceding analysis, I seriously question the veracity of almost all the corollaries of bolide impact (and extensive volcanism). The one exception is extensive darkness that would have severely curtailed photosynthesis in some but not all parts of the world. Even within the discussions of some of the strongest proponents of impact, I find doubt as to the severity of the "Dante's *Inferno*" corollaries. For example, in his concluding remarks in a 1994 article on paleosols at the K/T boundary in eastern Montana, Retallack (1994:1395) noted that "it is remarkable how little local soils changed following the disruption of plant and animal communities at the Cretaceous/Tertiary boundary. There are indications of crisis at the boundary, but they are slight and of shorter duration than the moderately developed paleosols at the boundary at Bug Creek." Later in the same paragraph he continues the same theme when he says "the striking extinction of plant and vertebrate species seem all out of proportion to paleoenvironmental changes indicated by paleosols." I am no expert on paleosols or fossil soils, so I will defer to others to judge what they can tell of past ecological perturbations. This might be especially problematic with Retallack's study, as it was limited to essentially one geological section in one part of the vast reaches of eastern Montana. Yet, if Retallack is correct in his assessment, it only adds to my analysis in showing that there could be no cataclysmic corollaries of impact, save extended darkness. This would give exactly the more subdued pattern of extinction and survival that Retallack found so remarkable.

In 1981, one year after the original impact paper by the Alvarez group, I wrote, "The paleontological data cannot be used to evaluate these hypotheses directly. However, it is only through paleontological investigation that the biological events at the Cretaceous-Tertiary boundary can be investigated." I have not been alone in emphasizing that physical evidence is required to test the efficacy of an impact (or flood-basalt volcanism). Paleontological data do not, or only poorly, address the issue of ultimate causes. Rather, such data are required in testing the magnitude and possibly the occurrence of the corollaries—the proximate causes.

The vertebrate data cast grave doubt on the presence and presumed killing effects of all the proximate corollaries of impact and volcanism (or at least their presumed intensities), except for the original proposal of extended darkness. It is my contention throughout this book that we must indeed listen to what the fossils say.

VIII

Left High and Dry: Marine Regression and Some Sundry Causes

Throughout geological history many areas of the modern terrestrial realm were inundated by shallow epicontinental seas. The term refers to their occurrence upon the continental shelves and platforms rather than in deep ocean basins. Epicontinental seas reached depths of only 1,500 to 2,000 feet, very shallow relative to most large modern marine bodies. Compared to the geological past, epicontinental seas are almost nonexistent today. Hudson Bay is the chief exception in North America. During the Late Cretaceous large areas of modern continents were submerged under warm, shallow epicontinental seas. Only recently has it become clear just how dramatic the loss of these seas was leading up to the K/T boundary.

Since the acceptance that continents are and have been on the move, atlases have been produced showing hypothesized continental positions through time (e.g, Smith et al. 1981). Knowledge of past continental positions was a boon to biologists and paleontologists who wished to make sense of animal and plant distributions. The older atlases, however, paid little attention to the presence and distribution of shallow seas. A group of British earth scientists (Smith et al. 1994) recently published a fascinating atlas showing not only the positions of continents in the past but also their coastlines (including epicontinental seas). The maps are sequenced at about eight million year intervals from the present all the way back to the beginning of the Mesozoic Era some 250 mya.

This new atlas vividly illustrates just how great the reduction of epicontinental seas was near the end of the Cretaceous. Figure 8.1 depicts this loss. The loss of shallow seas, and the consequent increase in nonmarine area, was greater surrounding the K/T boundary than at any other time in the past 250 million years. The nonmarine area increased from 42 million mi^2 to 53 million mi^2—more than a 25% gain. This is the equivalent of adding the land area of all of Africa. The second largest increase in continentality is also very apparent in figure 8.1. It occurs across the Triassic/Jurassic (T/J) boundary. As with the K/T transition, the T/J marine regression surrounds one of the five universally recognized mass extinctions during the Phanerozoic, the last 550 million years.

Figure 8.1 also depicts the sea-level curve by Haq et al. (1988). What

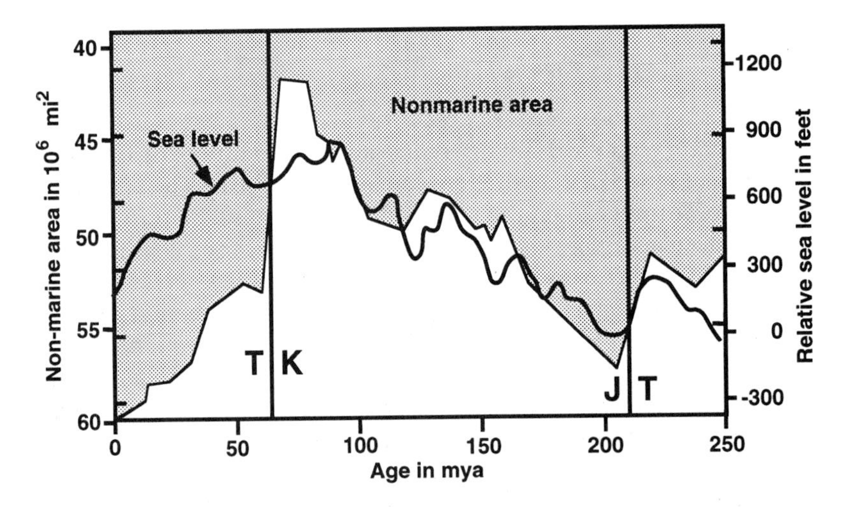

FIGURE 8.1 Sea-level changes. Changes in sea level (scaled on right) during the past 250 million years are juxtaposed with the calculated changes in the amount of continental area exposed to air (the nonmarine area; scaled on left). The two most abrupt large gains in terrestrial habitat bracket the K/T and T/J (Triassic/Jurassic) boundaries. That at the K/T, however, is not commensurate with the concurrent changes in sea level—probably because, as the epicontinental seas were drained, the flatness of the landscape meant large gains in terrestrial habitat for only small increments of sea-level drop. The sampling interval averages about ten million years. *Source: Sea-level changes after Haq et al. 1988; Nonmarine area after Smith et al. 1994.*

is of particular interest is the sea-level curve compared to the land area curve from mid Cretaceous (about 100 mya) to the Recent. After the K/T boundary, sea level dropped while land area increased. Two-thirds of the computed increase in land area seems to have occurred in only ten million years surrounding the K/T boundary. Although the disparity is noted by Smith et al. (1994), the authors do not attempt an explanation. I think the disparity in the two curves is in part real because the great tracts of extremely shallow seas lay upon areas of low continental relief. As sea level dropped even small amounts, large expanses of low-relief areas on the continents were exposed. This pattern is readily observed by comparing Smith et al.'s (1994) paleocoastline of the Late Cretaceous continents (figure 8.2) with that of the early Tertiary (figure 8.3). Even here, the early Tertiary reconstruction in figure 8.3 is about five million years after the K/T boundary, during a short-lived reincursion of seas (look at mid North America in figure 8.3). Closer to the K/T boundary the extent of epicontinental seas was even less.

Nowhere was this change more dramatic than in North America. Near the end of the Cretaceous, maximum transgression divided North America into two land masses (figure 8.2). As regression began and continued until at or near the K/T boundary, coastal plains diminished and became fragmented; stream systems multiplied and lengthened; and as sea level fell, land connections were established or reestablished (figure 8.3). Location of the shoreline during maximum

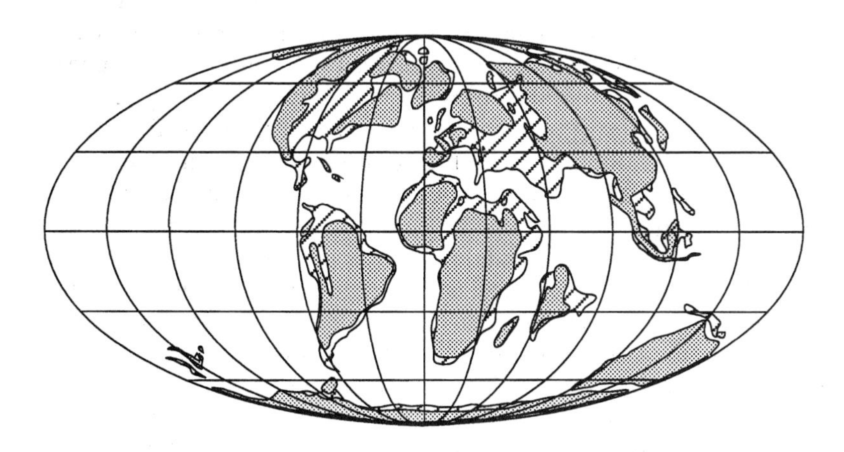

FIGURE 8.2 Paleocoastlines for the Maastrichtian stage of the latest Cretaceous, 70 mya. Nonmarine areas are shaded; epicontinental seas are hatched. After Smith et al. 1994.

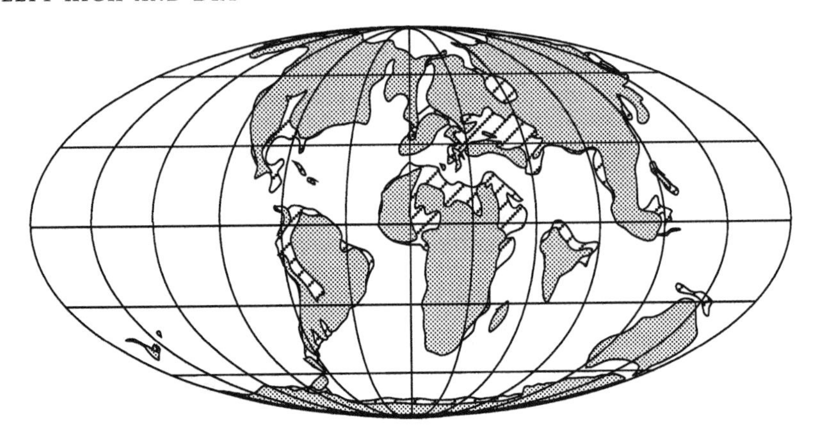

FIGURE 8.3 Paleocoastlines for the Thanetian-Danian stage boundary of the Paleocene, 60 mya. Nonmarine areas are shaded; epicontinental seas are hatched. After Smith et al. 1994.

transgression and its later shift during marine regression can be tracked during the latest Cretaceous in some areas of the Western Interior, such as the Dakotas, Montana, and Wyoming, using strand line positions indicated by marine invertebrates (e.g., Waage 1968; Gill and Cobban 1973). We know less about the very latest Cretaceous or earliest Paleocene maximum regression in the center of the Western Interior of North America, because the seas completely left the region, with the shoreline pushing south into Texas. Invertebrate and vertebrate marine species are absent from the earliest Paleocene in the Dakotas (Cvancara and Hoganson 1993). Such taxa reappear only briefly in the Dakotas during the short-lived, last-gasp reincursion of the area during the latest early or middle Paleocene (figure 8.3).

What drives these inundations, or transgressions, of the low-lying portions of continents is still not fully understood. The general consensus is that it is related to plate tectonics. In a recent paper modeling the process, Michael Gurnis (1993) presented the view that the rise in sea levels and inundations began as the motions of the plates increased. As this occurred, the margins along which the colliding plates converged were subducted, pushed down into the earth. The marine inundation of shallow continental shelves and platforms thus occurred not because the volume of water increased (as it has during our own interglacial time) but because the low-lying portions of the continents became even lower. Presumably, marine regression owes to the reverse. Whatever the geophysical factors driving the process, the physical manifestation of marine regression, like impacts and volcanism, must be tested as an important ultimate cause of extinction. To do so, again, one must look to the corollaries.

There are important physical and biological proximate corollaries to marine regression, some of which I have already alluded to in this brief introduction and which I now discuss further. I will also suggest that, as with marine regressions, transgressions have a role to play in speciation and extinction. In the next three sections I discuss, in turn, the three main proximate causes of extinction likely to attend a marine regression: habitat fragmentation, the lengthening of streams, and competition wrought by invaders that gain access by land bridges.

BREAKING UP IS HARD TO DO

Although marine transgression and regression near the K/T boundary was global in extent, I turn to a closer examination of how those changes played out in North America because this is where we have the vertebrate data at the K/T boundary. Figure 8.4 shows North

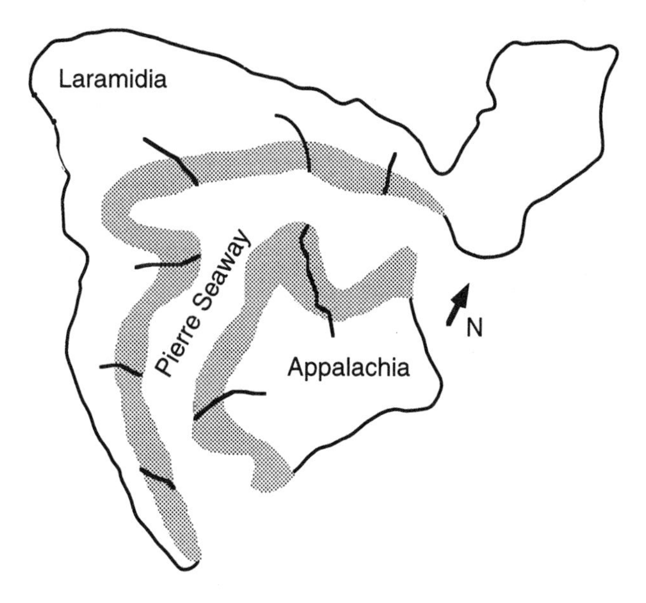

FIGURE 8.4 Coastal continuity during marine transgression. During the peak of the last marine transgression of the Late Cretaceous, about 70 mya, North America was bisected by the shallow Pierre Seaway into two land masses: Laramidia in the west and Appalachia in the east. Shaded areas are approximations of coastal areas bordering the seaway; lines are suggestive of streams that flowed into the seaway. After various sources.

America near the time of the last maximum transgression in the Late Cretaceous.

For almost forty million years during the Late Cretaceous, North America was split into a western and eastern portion by an epicontinental sea. This sea is given various names during its long existence. For the sake of brevity and clarity I shall call it *Pierre Seaway*, after a formation of the same name exposed in a number of western states. This is the name used in the classic atlas on the geology of the Rocky Mountains (McGookey et al. 1972). To avoid mental gyrations caused by phrases such as the east coast of western North America, I will refer to the land mass west of the Pierre Seaway as *Laramidia*, after the mountain-building episode that occurred during this time and later in western North America. The land mass to the east of the Seaway will be *Appalachia*, after the much older mountain range in that region.

Most of the latest Cretaceous vertebrate fossils come from the coastal region on the east coast of Laramidia. The west coast of Appalachia, as well as the eastern seaboard of Appalachia, have also produced some specimens. The exact extent of the coastal regions bordering the Pierre Seaway are conjectural, but the shaded regions in figure 8.4 provide a sense of just how extensive they were. These are the lands that supported a terrestrial and shoreline vertebrate community in a zone of deposition—and hence in a zone that could be preserved in the rock record. Similarly, we have only a limited sense of where actual stream systems existed; thus the lines shown on figure 8.4 serve only as a reminder that these systems were present. Stream systems are important for reconstructing the past because they hosted many of the vertebrate species.

In the last few million years of the Cretaceous the Pierre Seaway began to regress from both Laramidia and Appalachia. At or just shortly before the K/T boundary, the seaway reached its nadir. This retrenchment is shown in figure 8.5. The receding coastlines both north and south have not been well established, but we do know that the southern coastline reached well into Texas. Whatever the exact coastline, a dramatic reduction in coastal plains occurred at that time (compare the shrinkage in shaded area from figure 8.4 to 8.5). It is from this environment that we necessarily must sample the last of the Late Cretaceous vertebrate community.

A common refrain is that because total amount of land increased with the regression, dinosaurs should have had more, not less, area and more environments in which to live. We know with considerable certainty that dinosaurs did live in noncoastal environments, such as

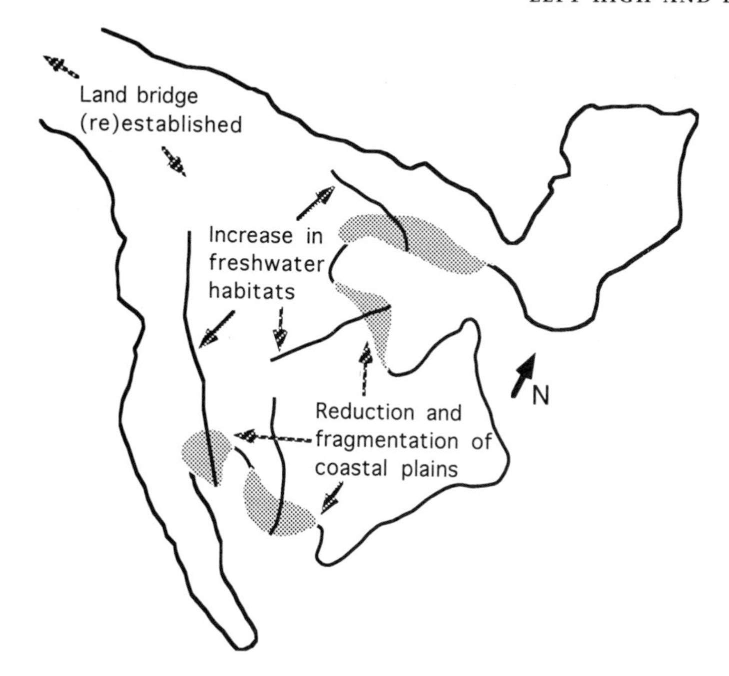

FIGURE 8.5 Coastal fragmentation and land bridging during marine regression. In the very latest Cretaceous, marine regression nearly drained the shallow Pierre Seaway. Shaded areas suggest the small remnants of coastal areas; lines are suggestive of greatly lengthened streams. Notice the land bridge established with Asia. After various sources.

the higher, drier Gobi Desert in Mongolia during part of the Late Cretaceous. At present, however, the only well-known vertebrate communities that preserve dinosaurs at the K/T boundary are coastal. Thus arguments about what dinosaurs and other vertebrates may or may not have been doing in other environments are moot. It is simply incorrect to say that the dinosaurs and other vertebrates may have survived elsewhere, when we have little or no information about other environments. We must test our theories using the data we have, not the data we wish we had.

The drastic reduction of coastal plains put tremendous pressure on some, especially the large, vertebrate species. Reduction of habitat—for example, in the Rift Valley System of modern East Africa—first impacts larger vertebrates (the mammals), as these tend to have the greatest habitat requirements per animal. In the shrinking coastal plains of latest Cretaceous North America the equivalent large vertebrates first affected were the dinosaurs.

A compounding problem of habitat loss, whether in East Africa today or the coastal plains of latest Cretaceous North America, is the fragmenting of remaining habitat. This process, as a result of human activity, has become known by ecologists as habitat fragmentation. Figure 8.6 illustrates the linkage between fragmentation and extinction. In larger, undisturbed habitats, animals (and plants) can move more easily from one area to another. If the habitat is fragmented, however, and even if the areal loss of habitat is not in itself substantial, the flow of populations from one fragment to another will be hindered. Habitat fragmentation is an underrecognized causal agent in natural extinctions in the paleontological past. I am joined in this view by Lev Nessov and Lena Golovneva, who, in their 1990 review of floral and faunal changes during the Late Cretaceous in the northeastern reaches of Siberia, express almost the same ideas.

For some species even seemingly small barriers, such as two-lane roads, can be insurmountable. The results can be disastrous if viable populations cannot be maintained in the various fragments. As figure 8.6 suggests, fragmentation can lead to extinctions. Barriers also arise in nature even among animals that would seem easily capable of dispersing. Although I have no doubt that the result is very often extinction, we usually only see what survives in the form of differences between closely related species. Primates, the group to which we belong, probably show this tendency. For example, thirteen subspecies of the saddle-backed tamarin (a squirrel-sized monkey) of western South America are found in adjoining areas separated by major rivers of the upper Amazon Basin. The tamarin subspecies are distinguished by often sharply defined differences in coloration and color pattern (Hershkovitz 1977; Albrecht and Miller 1993). Another example is the Kaibab squirrel on the North Rim of the Grand Canyon. Unlike its nearest relative, Abert's squirrel, which is found on the south side of the Grand Canyon and elsewhere in the western states and Mexico, the Kaibab squirrel is restricted to an area of only eight hundred square miles (Hall and Kelson 1959). Fragmentation, in this case the development of the Grand Canyon, helped produce the differences, but the margin between evolutionary origin and oblivion for the Kaibab squirrel has surely been slim.

Some earth scientists have suggested that habitat fragmentation is vague (Buffetaut 1994) or even untestable in the geological record (Fastovsky and Sheehan 1994) in the context of marine regression. Although habitat fragmentation may not be a well-understood phenomenon among earth scientists, it is all too real a phenomenon for biologists studying the effects of human activity in modern rain-

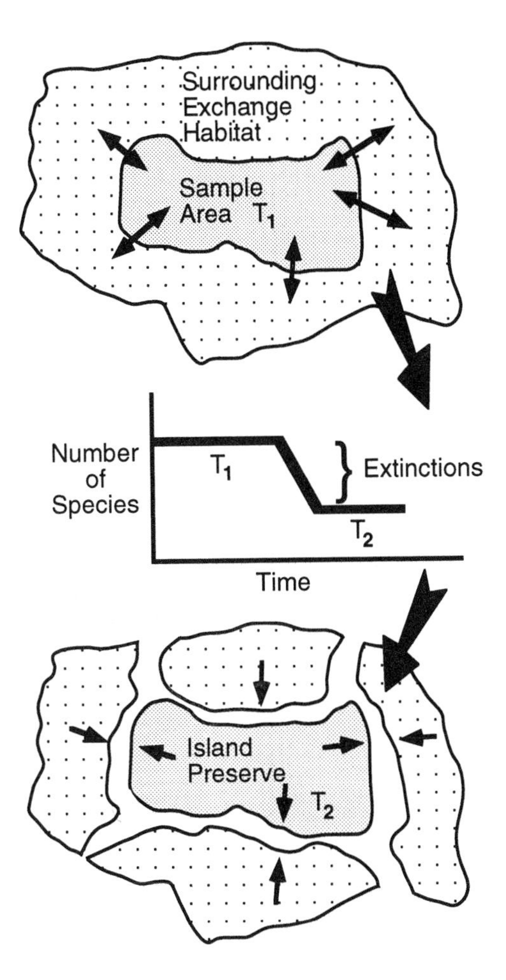

FIGURE 8.6 Habitat fragmentation as a cause of extinction. *Top:* In larger habitats, animals (and plants) move freely from one area to another. *Bottom:* Habitat fragmentation, even without much loss in aggregate area, reduces the flow of organisms from one fragment to another. The result is extinction and a shift to some new, lower equilibrium number of taxa. After Miller 1978.

forests and in urban settings. For example, Michael Soulé and colleagues (e.g., Bolger et al. 1991) have been documenting the decline of birds and mammals in San Diego, as urban development divides and isolates habitats in canyon areas. One would not expect that the natural equivalent of habitat fragmentation would be easily, if at all, preserved in the rock. The forcing factor for habitat fragmentation in the latest Cretaceous—marine regression—is, however, a thoroughly documented fact in the rock record of North America. Globally, marine regression occurred within this same general time frame, although how close in time it occurred in various regions is a matter of debate.

Recall that in table 6.1 I predicted large species would be the most severely affected by habitat fragmentation, for the reasons just discussed. As calculated from the data in table 5.1, only eight of thirty large species of vertebrates preserved in the rocks of Montana's Hell Creek Formation of the uppermost Cretaceous survive into the Tertiary. All eight were aquatic or at least partially aquatic (two fish, one turtle, one champsosaur, four crocodilians). In contrast, twenty-two large terrestrial species and one aquatic species became extinct. The tally of losers includes one turtle, one lizard, one crocodilian, and nineteen dinosaurs. Predictions of which of the twelve major taxa would be the most susceptible to extinction owing to habitat fragmentation (see table 6.1) fit the paleontological data quite well, with nine of the twelve agreeing in table 6.1.

UP A LAZY RIVER NO MORE

Extensive work has been done on the sedimentology and drainage patterns for some specific channel systems in the eastern part of Laramidia during the latest Cretaceous and early Tertiary (e.g., Belt 1993). But our understanding is far from complete. There is still no definitive map of drainage for all of Laramidia and Appalachia. Thus in figures 8.4 and 8.5 positions of streams are largely hypothetical. Even lacking such precision, we can be certain that as new land was added following marine regression in the early Tertiary, stream systems increased and lengthened. This process is the second major proximate corollary of marine regression.

A mitigating factor for these stream increases could be a general climatic drying, turning some formerly permanent streams into ephemeral flows and shrinking the size and number of bodies of standing water. In at least eastern Montana and western North Dakota the opposite occurred; the amount of ponded water increased across the K/T boundary (Fastovsky and McSweeney 1987).

Because freshwater habitats increased in the areas from which the epicontinental sea receded, most aquatic vertebrates did well across the K/T boundary, except for those with close marine ties—sharks and some bony fishes. Such fishes may need to spend at least a portion of their life in a marine environment, in some instances to reproduce. The major group most likely to suffer would have been the elasmobranchs. In fact, all five species of elasmobranch present in the Hell Creek Formation show no presence in the earliest Tertiary. It is not clear, however, whether these disappearances from the Western Interior are actually extinctions at the K/T boundary or whether the species survived elsewhere in marine environments into the earliest Paleocene. As yet, however, no such specimens have been reported.

Cvancara and Hoganson (1993) regard these elasmobranch disappearances as true extinctions. They show in a range chart that new Paleocene species of elasmobranchs appeared in the Western Interior at the K/T boundary. Their own data and discussion show, however, that this is not the case, as the definitively oldest marine sediments that postdate the K/T boundary in the Western Interior (the Cannonball Formation) are no older than late early Paleocene. This suggests a gap in marine sedimentation in the Western Interior of possibly one million years or more. This pattern of disappearance and reappearance of marine sediments strongly suggests that as the Pierre Seaway regressed further and further away from eastern Montana, all sharks and relatives departed, because connections to the sea became attenuated (figure 8.7). New species of elasmobranchs did not occur in the area until a smaller transgression reached this far inland at or just before middle Paleocene times. This is known as the Cannonball Sea, which is the more northerly portion of the seaway in North America, barely discernible in figure 8.3.

The total disappearance of sharks and relatives is the only prediction that can be made with any certainty to be an outcome of a loss of marine connections and the lengthening of stream systems. The increase of stream systems would have been a neutral or even positive factor helping to mitigate other stresses that may have been put on the freshwater system. Table 6.1 shows this prediction—and how it fares when the actual extinction data are compared. The fate of eight of the twelve major vertebrate taxa are in accord with this corollary.

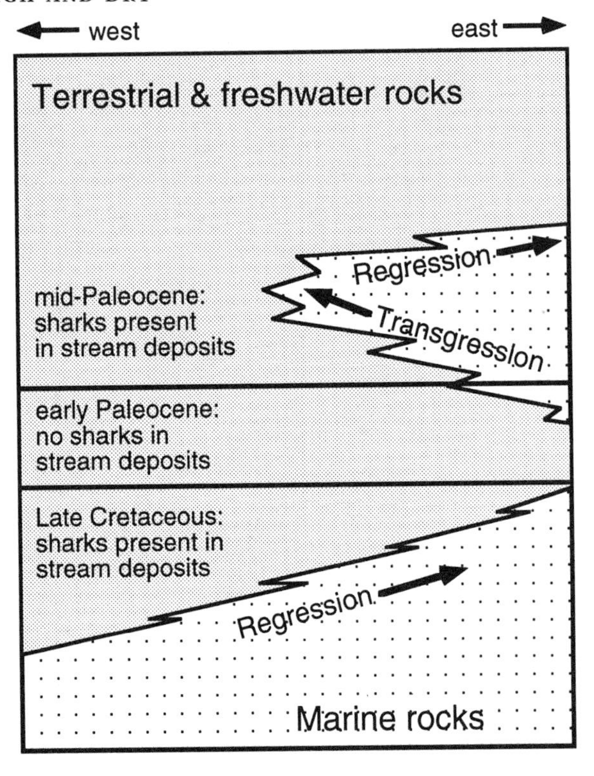

FIGURE 8.7 Effects of sea-level changes on riverine sharks. The presence or absence of elasmobranchs (sharks and relatives) in the latest Cretaceous and early Tertiary faunas of eastern Montana is determined by geographic proximity of marine conditions.

BRIDGE OVER TROUBLED WATERS

As sea level declined at the end of the Cretaceous, new land areas were exposed. In some cases this prompted the establishment or reestablishment of intercontinental connections. One such connection was the Bering Land Bridge (figure 8.5). At various times during the Late Cretaceous this bridge appeared and then disappeared. The pattern is suggested by similarities in parts of the Late Cretaceous vertebrate faunas in Asia and North America, especially the better-studied turtles, dinosaurs, and mammals.

The loss (or fragmentation) of coastal habitats and the increase in freshwater systems are corollaries about which we can make survival predictions for various groups, based upon physiology, size, habitat requirements, and so on. This is not the case for the establishment of land bridges and subsequent biotic mixing. Competition and then extinction often result from biotic mixing, but predicting the fates of various taxonomic groups is usually not possible.

As I discussed in chapter 5, the oldest marsupials are known from approximately 100 million year old sites in western North America. As figure 8.8 shows, by about 85 mya we know of about ten species of marsupial. This number rose and stayed at about fifteen species from about 75 mya until the K/T boundary 65 mya, when the marsupials plummeted to one species. All marsupials in the latest Cretaceous of eastern Montana were quite small, from the size of a mouse to that of a very well fed opossum or raccoon. Their teeth were much like that of modern opossums, with slicing crests and well-developed but relatively low cusps (compared to contemporary placentals) for poking holes in insect carapaces, seeds, or whatever they found. Most did not appear to be specialists on any particular food.

With the reestablishment of the Bering Land bridge (or at least closer islands) near the K/T boundary, a new wave of placental mam-

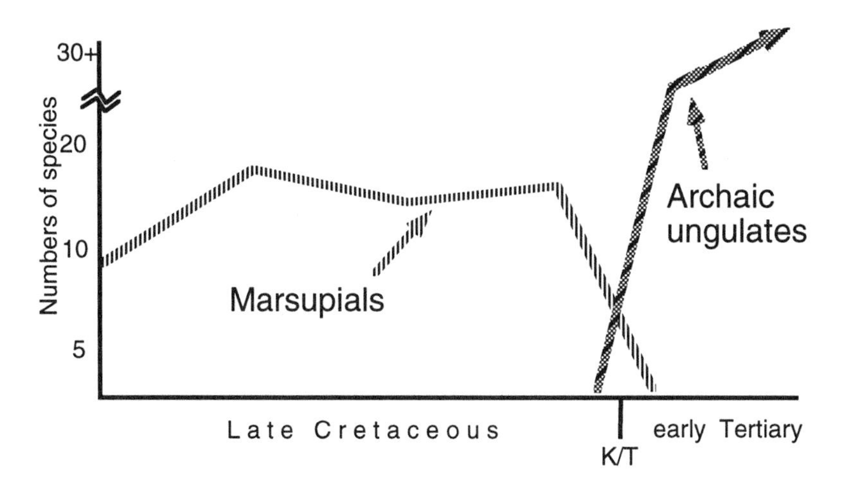

FIGURE 8.8 The rise of the ungulates and the fall of the marsupials in North America. By some 85 mya, about ten species of marsupial are known in North America. Marsupials increased to about fifteen species before their diversity plummeted to one species in the early Tertiary. New placental mammals, notably archaic ungulates, appeared right around the K/T boundary and began radiating in western North America coincident with the very rapid decline of marsupials. Within one million years of the K/T boundary, thirty species of archaic ungulates are known in North America, and numbers of species kept rising. After various sources.

mals appeared in western North America. These invaders (traditionally known as condylarths) were the very early relatives of modern ungulates and whales. As figure 8.8 shows, their appearance in North America coincides with the very rapid decline of marsupials. Within a million years of the K/T boundary thirty species of these archaic ungulates are known in North America, and their numbers kept on rising. Our best guess now is that the lineage that gave rise to these mammals first appeared in middle Asia 85–80 mya (Nessov 1987) and reached North America near the K/T boundary. Of special interest is that the archaic ungulate invaders had dentitions very similar to their marsupial contemporaries and presumably ate similar things. It seems more than coincidence that marsupials did well in North America for about twenty million years, only to almost disappear with the arrival of the ungulate clade. Consider, too, that both marsupials and ungulates were joint invaders of South America very soon after the K/T boundary. Their dentitions were then already beginning to show differentiation, with the marsupials headed toward carnivory and ungulates toward herbivory.

In table 6.1 I have listed the near total extinction of marsupials as the only predicted direct result of the reestablishment of a land bridge. Although the survival and extinction data somewhat support the prediction that flows from this corollary (eight out of twelve coincide), I again emphasize that, unlike the other two marine regression corollaries, a competition corollary cannot be considered as widely predictive.

In combination, the three corollaries of the marine regression theory provide substantial corroboration for this theory as an ultimate cause of end-Cretaceous extinction. Eleven of twelve extinction/survival patterns recorded in the rocks for major vertebrate taxa match the predictions. The one pattern that does not fit any of the predictions of the three general corollaries of the marine regression theory is for lizards, which have a high percentage of extinction (70% — 7/10 species). A more local or even regional change might be the culprit. As I have noted several times, following the K/T boundary in eastern Montana we see a raising of the water table, increased ponding of water, possibly increased rainfall, and most obviously coal deposition. This marks a shift from a drier to a wetter environment. If the lizards were mostly dry-adapted, this may have been the reason for their departure from the region and/or their total extinction. This must be regarded as a more local and ad hoc explanation. One could argue similarly that the effect on marsupials, which flows from the competition corollary, should not be included because it is a specific, not a general prediction—in which case the agreement drops to ten of the twelve extinction/survival patterns. Overall, the predictions suggested that regression would be hardest on terrestrial species, especially those of larger size, while aquatic forms would fare better by comparison. This prediction was indeed borne out by the data.

Thus far, marine regression as a possible cause of major extinction looks robust. By this analysis, its predicted effects show a tighter fit with the paleontological data than do the impact or volcanic theories. A next step would be to ask if the opposite of marine regression—namely, transgression—would have an opposite affect on a vertebrate biota. Do freshwater species suffer greater levels of extinction and do terrestrial species fare better during transgression?

THE SEAS COME IN, THE SEAS GO OUT

During most of the last forty or so million years of the Cretaceous, North America was divided into a western Laramidia and an eastern Appalachia, although the midcontinental seaway (under various names) did wax and wane. In his authoritative book on sea-level changes throughout the Phanerozoic, Tony Hallam (1992) indicates four major transgressive-regressive cycles in North America during the Late Cretaceous through the earliest Tertiary. Figure 8.9 is adapted from Hallam's book.

With some notable exceptions, most of the Late Cretaceous record of vertebrates, including dinosaurs, comes from the regressive phases. These represent times when more terrestrially derived sediments from streams, swamps, and lakes are formed and preserved as the seaway retreated. When more of these kinds of sediments are preserved, more bones are preserved as well. Most of our knowledge of North American Late Cretaceous vertebrates is concentrated in the last two regressive phases, especially the well-known Judith River faunas in southern Alberta and Montana and the Hell Creek–Lance faunas of Montana and Wyoming. We only have an inkling of possible trends relating patterns of vertebrate evolution and sea-level changes. Two patterns occur during transgressions, and a third is the one I have been describing during the last regression in and beyond the Cretaceous. These patterns hint at an answer to the question I posed at the end of the last section; freshwater species do seem to suffer more extinction and terrestrial species do seem to fare better during transgression.

In southern Utah, paleontologists Jeff Eaton and Jim Kirkland have been studying the earliest reported maximum transgression in the Late Cretaceous. This transgression occurs across what is known as

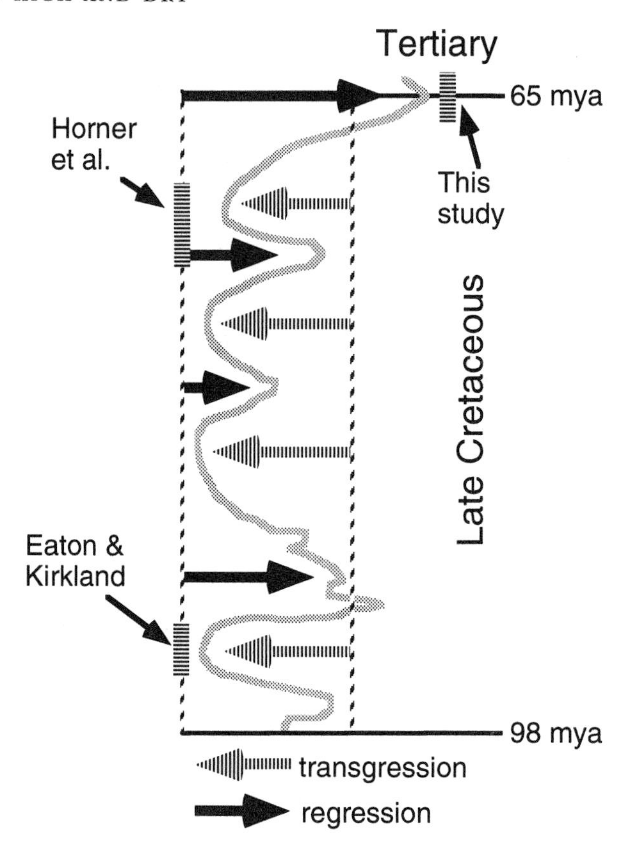

Figure 8.9 The four major transgressive-regressive cycles in North America during the Late Cretaceous. The first marine transgression (studied by Eaton and Kirkland) occurs at the Cenomanian/Turonian boundary. Horner and colleagues studied the final transgression. In earlier chapters here I presented the results of my own study bracketing the K/T boundary. *Source: The sea-level cycle depicted here is after Hallam 1992.*

the Cenomanian/Turonian boundary; it is marked as "Eaton & Kirkland" in figure 8.9. In the past few years I have sometimes accompanied Jeff Eaton in his field endeavors. The fossil sites can be few and far between, but the beautiful geological terrain and cooler mountain air more than compensates. What Eaton and Kirkland found is that near maximum transgression, the riverine-riparian vertebrates show notable (probably true) extinction. Meanwhile, groups of terrestrial vertebrates continued to radiate (Eaton and Kirkland 1993). One of the more interesting extinctions at this time affects the lungfishes.

Species of lungfish are today still extant in Africa, Australia, and South America, but this freshwater group was forced to extinction in North America during the first maximum transgression in the Late Cretaceous (Kirkland 1987). Even more recently, Eaton (written comm. 1994) reported that during this transgression, "there is no extinction 'event' recorded in the brackish water faunas. Freshwater and riparian faunas were strongly affected as indicated by the extinction of some turtles, crocodilians, and fish. Fully terrestrial organisms such as mammals and nonavian dinosaurs appear to diversify across the boundary."

Moving forward in time, we come to the penultimate regression and ultimate transgression in the Late Cretaceous. Jack Horner and colleagues studied the evolutionary history of some of the dinosaurs at that time in the Western Interior (1992). They argued that during the five million year deposition of the Judith River Formation, there was little change in dinosaurs and mammals.

Like the Hell Creek Formation, the Judith River Formation has yielded a notable vertebrate fauna, especially rich in dinosaurs. Also like the Hell Creek fauna, the Judith River fauna developed on the coastal plains during a marine regression. There is, however, a major difference: the Hell Creek regression was far more extensive (see figure 8.9). By the end Cretaceous the midcontinental sea had all but disappeared, bringing about a tremendous reduction and fragmentation of the coastal plains. This degree of change did not occur during Judith River times.

In Judith River times, an expansive coastal plain developed during regression and remained (unlike during later Hell Creek times) because a smaller midcontinental sea still bisected North America. The evolutionary stasis reported by Horner and his colleagues (1992) for dinosaurs and mammals occurred during the maximum Judithian regression, which reduced but did not eliminate the midcontinental seaway. Horner and colleagues documented anagenetic appearances within four separate lineages of dinosaurs coincident with the beginning of the last Late Cretaceous marine transgression, which advanced the coastline to the west during the waning years of Judith River time. Anagenesis is evolutionary change in an existing lineage without the splitting (or cladogenesis) into more lineages. Thus transgression appears to have been strong enough to force some evolutionary change, but not to create whole new lineages or cause major extinctions.

The study performed by Eaton and Kirkland at the Cenomanian-Turonian boundary, the study performed by Horner and colleagues in strata deposited later in the Cretaceous, and my own study that brack-

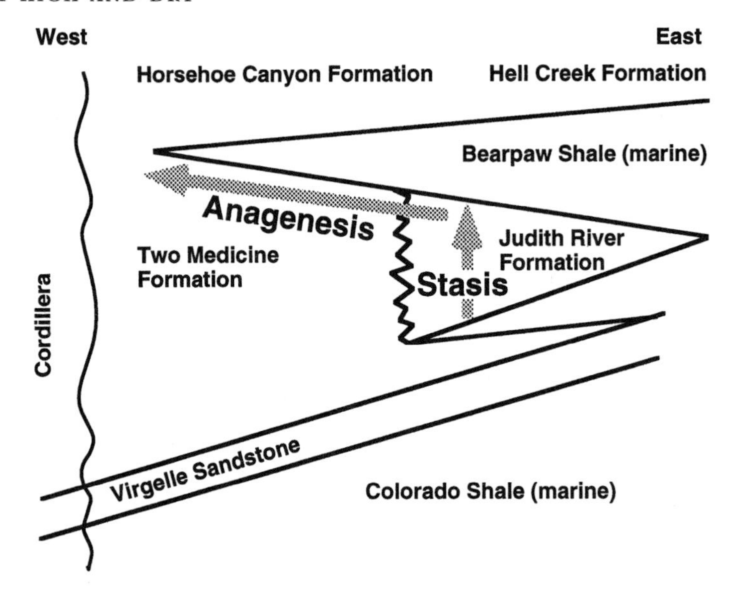

FIGURE 8.10 Evolutionary stasis v. anagenesis during sea-level changes. During the five million years in which the Judith River Formation was deposited (marking the penultimate regression of the epicontinental sea in the Late Cretaceous), there was little change in dinosaurs and mammals. Evolutionary stasis came to an end as the ultimate Late Cretaceous transgression began. That transgression coincides with significant evolutionary change (anagenesis) in four major clades of dinosaurs, recorded in the uppermost Judith River Formation and the Two Medicine Formation further to the west. After Horner et al. 1992.

ets the K/T boundary all point to the same conclusions. Overall, it appears that marine regressions are hardest on terrestrial (especially large) vertebrate species, but that freshwater forms are favored during such times. Conversely, transgressions are hardest on freshwater species but favor fully terrestrial forms. These three studies are only suggestive of the role that these cycles may play in vertebrate evolution and extinction. The evidence is, however, strong enough to suggest that transgression and regression may be forcing factors in the evolution of at least the vertebrates.

COROLLARIES WITHOUT A CAUSE

The corollaries I have just discussed (habitat fragmentation, stream lengthening, competition), as well as those in the previous chapter (sudden cooling, acid rain, global wildfire), would seem to be attributable to specific ultimate causes. Two additional corollaries may have accompanied any one of the three ultimate causes—that is, they might have been induced by impact, volcanism, or marine regression. Thus these two are "corollaries without a cause." By this I mean that they are the result of a combination of two or more ultimate causes.

Hot Time in the Old Town: Local Fires

The evidence of extinction and survival patterns for vertebrates presented in table 6.1 is quite compelling in showing that a global conflagration at the K/T boundary is a highly unlikely explanation. Nevertheless, the same patterns do seem consistent with more localized fires. Fully terrestrial species would be more adversely affected by local wildfires than would partially or fully aquatic species, and thus there is agreement in nine of the twelve extinction/survival patterns (table 6.1).

An increase in local fires might accompany the drying and increased plant mortality as habitats and climates shifted during marine regression and during prolonged volcanic eruptions. If plant communities were more vulnerable to fire, an earth-shrouding impact causing a dramatic decrease in or cessation of photosynthesis in some regions would only accelerate the drying and death of plants. Just as in a drought, the vulnerability of plants to fire would have increased, but on a much grander scale. This kind of scenario, based upon our knowledge of physical events (impact, volcanism, and marine regression), addresses very well the pattern of plant survival and extinction at the K/T boundary. A problem for this local effect in eastern Montana, however, is that, as I have noted, the climate seems to have become wetter. I comment further on this possible effect of local fires in the closing chapter.

Down and Dirty: The Detrital Influx Hypothesis

One result of increased plant mortality, whether fire was a primary culprit or not, may have been an influx of plant detritus into freshwater systems. This detrital influx hypothesis was introduced as a corollary of impact by invertebrate paleontologist Peter Sheehan and sedimentologist David Fastovsky (1992). They argued that detritus-based feeding in the freshwater realm versus dependence on primary production on land allowed much higher survival of freshwater versus land-dwelling species. (Detritus is bits and pieces of organic matter, from microscopic to log-size; primary production refers to material produced by green plants.) Sheehan and Fastovsky came to this explanation for the preferential survival of freshwater species compared to land species when they reexamined the 1990 study by Laurie Bryant and me. They argued that there was a 90% survival rate for freshwater species versus only 10% for land-dwelling species.

For the sake of brevity I will call their idea the *detritus-feeding corollary*. In an important sense, this corollary is different from those already discussed (except perhaps competition). This is because the corollaries of a sudden cooling, acid rain, and global wildfire, or of habitat fragmentation and stream lengthening, are hypothesized as *direct* physical consequences of an impact, volcanism, or marine regression—whereas detritus feeding is a biological result one step removed from such physical events. Thus if an impact is posited as the first-level (ultimate) cause of the vertebrate survival/extinction pattern at the K/T boundary, a global wildfire would be a second-level (proximate) cause, and an influx of detritus into water bodies would be third level.

One can construct a wide array of possible scenarios for what happens to primary and detritus production following an impact. Let us begin with what is certainly the worst-case scenario: the impact is massive, acid rain pummels the Earth, temperatures plummet, and global wildfires torch much of the above-ground biomass. Lest this seem an exaggeration, recall the conclusion drawn by Walter Alvarez that the K/T impact was a veritable "Dante's Inferno of appalling environmental disturbances" (1986:653) and Alan Hildebrand's equally odious image: "The K/T impact turned the Earth's surface into a living hell, a dark, burning, sulfurous world where all the rules governing survival of the fittest changed in minutes" (1993:122). In these scenarios, a great percentage of land plants would die, with a resultant precipitous decline of primary production on land. All the burned and half-burned material would pour into freshwater systems and thence into the oceans.

This does not mean that we should expect to find in the rock record large accumulations of broiled dinosaurs or other animals where they were stopped in their tracks or of plants where they were rooted. This is because most terrestrial animals and plants lived and died in areas not conducive to preservation (e.g., see Cutler and Behrensmeyer 1994). If the idea of mass killing, especially global wildfires were true, however, we should be able to see the detrital residue in most terrestrially derived sediments that accumulated at the K/T boundary. Also, the land would be denuded of plants to such an extent that we should see tremendously increased erosion and a noticeable influx of sediment in terrestrial K/T boundary sections.

No such concentrations have ever been recognized, however. Most K/T boundary sections in terrestrial rocks are thin, often carbonaceous units showing no great influx of organic or inorganic material. In fact, these finer sediments are the major reason that iridium has been recognizable in such terrestrial K/T sections. One possible exception that comes to mind is the Bug Creek channel sequence discussed at length in chapter 3. This extensive channel deposit, which was originally thought to be latest Cretaceous in age, is now believed by most detractors and supporters of a catastrophic impact to postdate the K/T boundary. Thus, at least for the worst-case scenario, the detritus-feeding corollary has no physical evidence supporting it.

One could imagine a much less drastic set of events leading to increased detritus feeding—with or without an impact. If my assessments are correct about what the vertebrate fossil record across the K/T boundary says regarding sudden cooling, acid rain, and global wildfire, these three corollaries did not occur or occurred in a much subdued way. Although Sheehan and Fastovsky (1993; Fastovsky and Sheehan 1994) are not explicit in saying just how bad things may have become, they do conclude that reduction in primary production was the result of an impact. It appears that they emphasize reduction of primary production rather than increase in available detritus. In my view, however, the two go hand in hand. The problem bedeviling waste management today serves as a useful analogy. As we produce more garbage (dead and burned plants, in the worst-case scenario of impact) we must find somewhere to put them (the detritus from the dead and burned plants).

As Sheehan and Fastovsky (1992) pointed out, the ecologies of riverine systems can vary considerably in their energy sources derived from primary production versus detritus. Peter Sheehan, speaking at the 1994 Snowbird III meetings in Houston, reported that detrital feeding is of great importance in streams, citing the work of Closs and Lake (1993). Closs and Lake, however, studied an upland, intermittently flowing stream, whereas those flowing through eastern Montana during the latest Cretaceous are thought to have been lowland, permanently flowing streams. Thus, just how strongly one can draw parallels between the two kinds of streams remains a question.

As Vannote et al. (1980) noted in their classic paper on river systems, primary production in streams is generally less important in the feeder streams near the headwaters, increases toward the midsection of streams, and then decreases in larger streams or rivers, owing to depth and turbidity. The waterways that flowed through eastern Montana in the latest Cretaceous were mid to larger size streams. Thus Sheehan and Fastovsky (1992) may well be correct that detrital feeding was important in the ecologies of Late Cretaceous waterways recorded in the rocks of Montana. More controversial is their claim that detrital feeding would have differentially favored aquatic species over terrestrial species on a global scale during a disaster such as thought to accompany a major impact. In my view, however, the pattern recorded in the rocks could just as well be explained by marine regression. As I already discussed, a proliferation and lengthening of riverine systems would accompany a marine regression, and thus could also have favored freshwater systems.

Sheehan and Fastovsky's (1992) calculations of patterns of survival and extinction in terrestrial versus aquatic taxa present a more extreme view of differences than do the calculations I presented in figure 6.2. They found a 90% survival for freshwater species and only 12% for terrestrial species; my own calculations were 78% and 28%, respectively. Sheehan and Fastovsky's more extreme figures were derived from some understandable—and some not so understandable—corrections of Bryant's and my original data set for latest Cretaceous vertebrates in eastern Montana (Archibald and Bryant 1990). One of the understandable corrections was their exclusion of rare species. As I described in chapter 6, the rarity issue is thorny.

A less understandable correction, however, was their treatment of all mammal disappearances as true extinctions, with no accommodation of possible speciation. According to their calculations, mammals fared no better than the dinosaurs at the K/T boundary as recorded in eastern Montana. Hell Creek mammal species suffered total extinction at the K/T boundary. Mammals of course must have survived somewhere else, later recolonizing the world. I find this NIMBY ("not in my backyard") view of evolution weak, especially because Sheehan and Fastovsky argue that the physical conditions driving the extinctions in Montana were global in effect. Instead we should evaluate which of the latest Cretaceous mammal species recorded in eastern Montana have very close descendants in the early Tertiary. For these species we should judge disappearances only as pseudoextinction—not true extinction.

Finally, I take issue with Sheehan and Fastovsky's decision to exclude from their analysis all five species of shark (or shark relatives) recorded in the Hell Creek, and which Bryant and I had included in our own analysis. Their rationale was that the sharks were not strictly freshwater. As I discussed in my response (1993b) to their paper, and as Bryant had already noted in her 1989 monograph, the sharks were not alone in their saltwater adaptations; six of thirteen bony fishes we then recognized in the aquatic fauna of the Hell Creek Formation were also not strictly freshwater. Such fishes either frequent coastal waters or spend some part of their life cycle at sea, much as salmon do today. The sharks and relatives were just as much a part of the freshwater ecosystem of latest Cretaceous eastern Montana as were the frogs and crocodiles. Correcting their analysis just for this factor, brings the boundary survivors to 78% (36 of 46) for freshwater species and 28% (16 of 57) for land-dwelling species.

I agree with Sheehan and Fastovsky that the paleontological record in eastern Montana reveals disparate levels of extinction between freshwater and terrestrial realms. Table 6.1 indicates that nine of twelve major taxa demonstrate an accord with the extinction-survival patterns predicted by the detrital influx hypothesis of Sheehan and Fastovsky. I do not, however, agree with these authors that the aquatic-terrestrial distinction is the sole pattern of importance at the K/T boundary. In the other two purported mass extinctions for which data are available for both terrestrial and aquatic vertebrates (Permian/Triassic and Triassic/Jurassic), aquatic vertebrates also fare better (Bakker 1977; Padian and Clemens 1985). In our 1990 paper, Bryant and I emphasized the differential nature of the K/T extinctions, even using this adjective in the title. We concluded that there is no single overriding pattern of survival and extinction that clinches an argument for one or another ultimate cause of extinction.

Figure 8.11 shows the survival-extinction patterns for three broad distinctions among taxa, in addition to the aquatic-nonaquatic distinction just discussed. These distinctions are ectothermic v. endothermic, small v. large, and nonamniote v. amniote. All are based on my most recent species counts, as given in table 5.1. Any or all four of these patterns may be the result of false correlations. I doubt this, however; I think that all four patterns should be probed for what they can tell us about events at the boundary. Here I use them to reiterate a central point that Bryant and I made in 1990: there is no single dominant pattern between surviving species and extinct species. Rather, there are multiple factors that determined whether a species did or did not survive, some of which I am sure have yet to be detected.

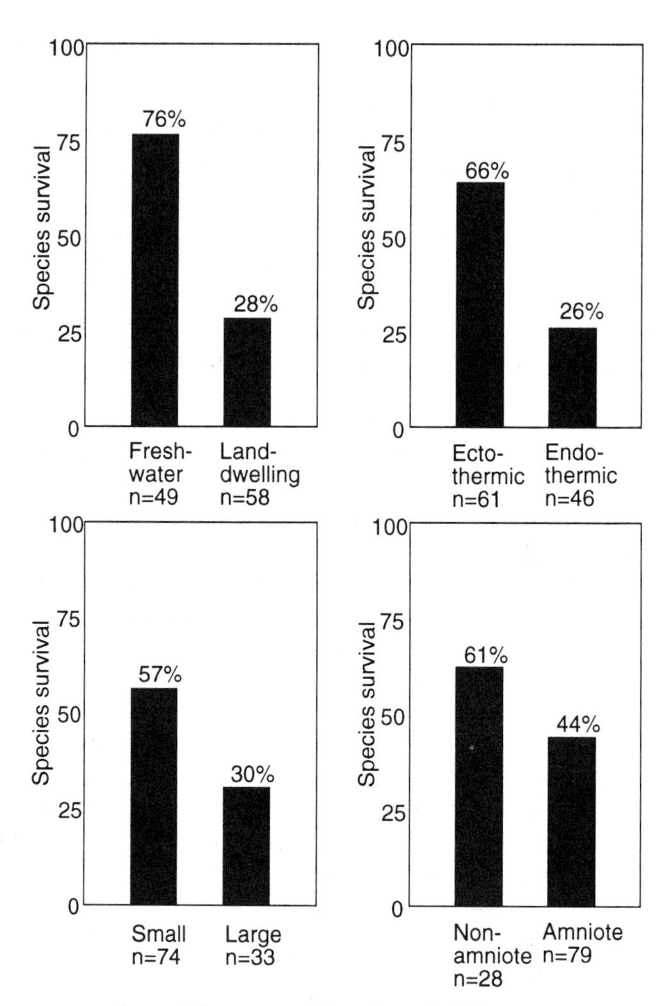

FIGURE 8.11 Biological factors correlated with different survival patterns. Vertebrate survival/extinction patterns at the K/T boundary appear to correlate with at least four biological factors: freshwater species fared better than land dwellers; ectotherms were favored over endotherms; small-bodied taxa (weighing less than about twenty-five pounds) did better than the large; nonamniotes did better than the amniotes. There was thus no single, overriding biological pattern for K/T boundary survivals and extinctions.

IX

Plants and Marine Organisms Across the K/T Boundary

I approach the mysteries of the K/T extinctions unrepentantly from the perspective of the vertebrate fossil record. These are the animals I know best. As a vertebrate paleontologist I am also painfully aware of instances in which scientists who are not vertebrate paleontologists fail to fully deal with or even misconstrue what the vertebrate fossils across the K/T boundary actually say—and, just as important, on what issues the fossils are silent. Theories of extinction must ultimately stand, fall, or at least be revised based upon how well they explain patterns of biotic turnover. Simple counts of the percentages of extinction or survival in various taxa do not constitute a test of a theory's veracity.

So far, I have gone through who the vertebrate players were, what their pattern of turnover was at the K/T boundary, and how well these patterns support corollaries of the impact, volcanism, and marine regression theories, plus two "corollaries without a cause": local wildfires and the detrital influx hypothesis. Is there a possibility that, as I have hinted, the K/T extinctions were not the outcome of a single horrific cause but rather the result of a cacophony of causes that ushered in the biotic changes that distinguish what has come to be known as the K/T boundary? I will explore this issue in the final chapter. But first I must recount, if all too briefly, the turnover patterns at the K/T boundary of taxa about which I am not an expert—notably, plants and marine invertebrates. I start with some of the most important contemporaries of the terrestrial vertebrates in the Western Interior, the plants.

PLANTS IN THE WESTERN INTERIOR

Plants, as they are studied in the fossil record, generally come in two sizes—megaflora and palynoflora. *Megaflora* entail wood and reproductive parts, but most often and probably most importantly, leaves. Palynofloral studies include the pollen of flowering plants, conifers, and cycads, and spores of everything else. These tiny and often mobile reproductive structures are collectively referred to as *palynomorphs*. Megafloral and palynofloral studies tend to be useful for answering different kinds of questions, and changes at the K/T boundary are no exception.

If the rock is right—that is, usually fine-grained sediments—palynomorphs can be found in prodigious numbers in just a fist-sized sample. The only downside for palynomorph collectors is that the fossils are usually invisible outside the lab. In general, because palynomorphs are structurally simpler than leaves, it can be more difficult to identify them to lower taxonomic levels. But their ubiquity makes up for this impediment. Palynomorphs can be invaluable in assessing overall floral change.

It can also be difficult to find fossil leaves, which are more like fossil vertebrates in that you tend to find them where they are rather than where you necessarily want them. As with palynomorphs, leaves tend to be better preserved in finer sediments, but I have seen exquisitely preserved leaf impressions even in sandstones. In fact, it is when leaves are in better indurated rock that I tend to stumble upon them. This is because they have been eroded from rock that is hard and well cemented, and thus more often preserves the plant remains. Localities preserving fossil vertebrates are also often first noticed in the same way. Some of the best plant localities I have ever seen, however, were discovered after a methodical search of finer rock units that have no visible remains at the surface.

I have watched one of the experts on latest Cretaceous and early Tertiary floras, Kirk Johnson, do his magic in finding fossil leaves in this persistent manner. Not surprisingly, he has made one of the most detailed analyses of floral change through the latest Cretaceous into the early Paleocene. As with the vertebrates I have discussed, Johnson's megafloral analysis is from the central portion of the Western Interior. Although he and I come to rather different conclusions about the importance of possible extinction mechanisms, I have been influenced by his papers and his discussions.

Megaflora

Kirk Johnson's fossil leaf collection from the Western Interior is impressive—nearly 25,000 specimens from some two hundred localities in eight areas from eastern Montana, eastern Wyoming, and the western portions of North and South Dakota (Johnson and Hickey 1990; Johnson 1992). His primary study area included about forty geological sections spread across five hundred square miles near Marmarth in southwestern North Dakota. In a composite section exceeding three hundred vertical feet, he recognized three major megafloral (leaf) zones (HC I, HC II, HC III), as well as some subzones in the upper Cretaceous Hell Creek Formation and one in the lower Paleocene (FU I), which in this area is called the Fort Union Formation (figure 9.1). The zones were distinguished primarily on the basis of dominant leaf types. This zonation is far more refined than we can recognize for vertebrate fossils. With the exception of the Bug Creek sequences, which I believe are now best regarded as Paleocene in age (see discussion in chapter 3), the vertebrate fossil record is not such that we can determine if there were any changes within the Hell Creek Formation.

Johnson's floral analysis included 247 taxa. At present, he is not able to assign formal names to all of the different leaf types in his collection; rather, he refers to them as *morphotypes*. Although he does not outright equate morphotypes with distinct species, for our purposes we can regard them as such. Thus, I will refer to them as "species" (in quotations) as a reminder that this designation was not the original usage.

Overall, Johnson discerned a 41% survival of "species" between zones HC I and HC II, a 25% survival between HC II and HC III, and a 21% survival between HC III and FU I. The last zonal boundary is also the K/T boundary. Contrast this with the almost 50% survival I found for vertebrate species across the K/T boundary. Phrased the other way round, he found almost 80% extinction, while I found slightly more than 50%. I can think of at least three, not mutually exclusive possibilities as to why terrestrial flora and fauna might show such different numbers.

First, perhaps the analyses of plants and vertebrates were done differently. Recall from the discussion in chapter 6 that I included both rare and common vertebrate species, excluding only eight problematic

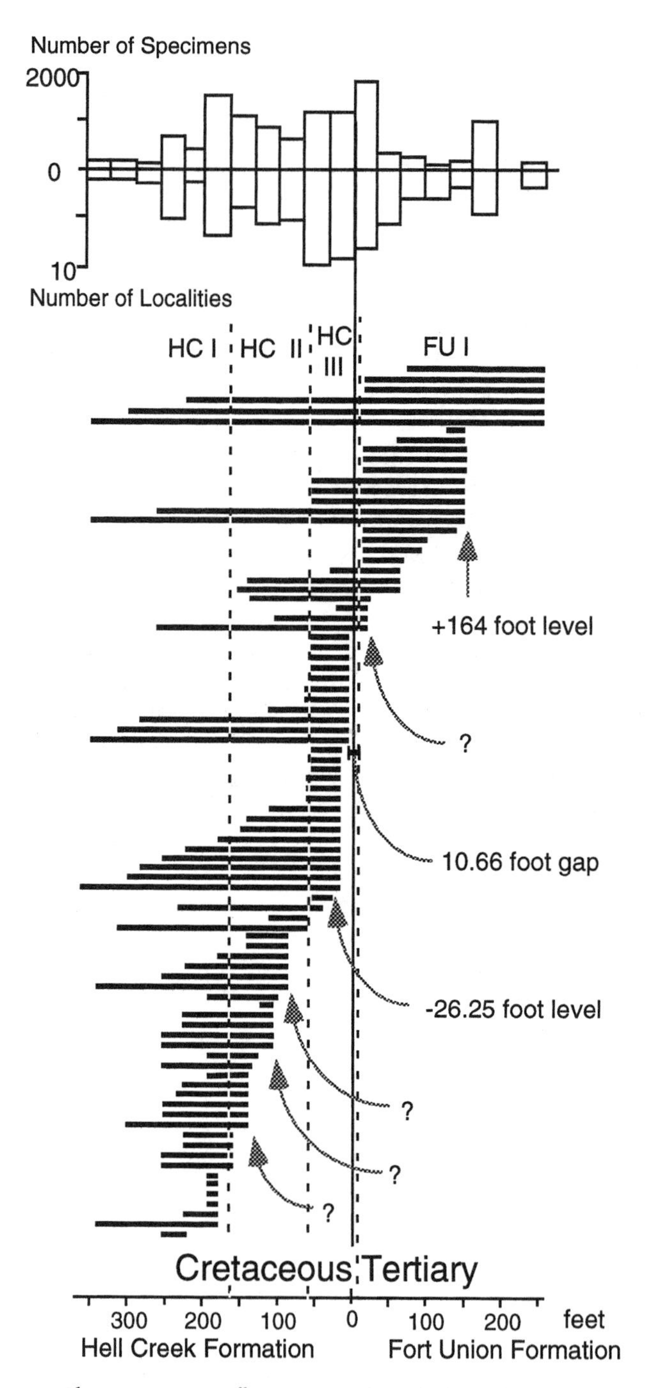

FIGURE 9.1 Changes in megaflora across the K/T boundary. Temporal ranges of fossil leaf "species" are based on fossil data from about forty geological sections of the Hell Creek Formation spread over five hundred square miles near Marmarth, southwestern North Dakota. The composite section for the Late Cretaceous extends through more than three hundred vertical feet. In it are three major megafloral (leaf) zones (HC I, HC II, HC III). The lower Paleocene Fort Union Formation contains just one recognized megaflora (FU I), which extends for more than two hundred vertical feet. Gaps and floral changes are indicated by arrows. After Johnson and Hickey 1990.

species used in the original analysis that Bryant and I published in 1990. It appears that Kirk also used all recorded "species" in his analysis. Thus different treatment of rarity is probably not a point of departure between the two studies. We must look for another cause.

Second, recall that I included as survivors those vertebrate species known to survive the K/T boundary outside the study area of eastern Montana and those species lineages that appear to have survived the K/T boundary in the form of descendant species (pseudoextinction). The plant study does not yet appear to have the geographic or, more important, systematic sophistication to detect these artifacts. The geographic limitation could probably be easily remedied because the areal extent of documented megafloras at the K/T boundary is better than that for vertebrates. In response to my concerns, however, Johnson recently told me that the second of these two possible artifacts, pseudoextinction, was probably not a significant problem in his analysis.

A third possible reason for the disparity in levels of survival/extinction between vertebrates and plants is that the disparity is not an artifact of any kind. Rather, the disparity may be real. There is, after all, no a priori reason why plants and vertebrates should be expected to fare exactly the same in the same circumstances of environmental stress. The levels of extinction were just as differential between plants and vertebrates as between various groups of vertebrates.

I can only hazard a guess as to the relative importance of each of these three factors. If Johnson's hunch is correct, pseudoextinction is not a problem. Some portion of the disappearances in plants may owe to local extirpation rather than total extinction, but in Johnson's view the extinction percentages he has already calculated would not be changed by much. This means that the different levels of extinction of plants and vertebrates (80% versus 50%, respectively) in the Western Interior at the K/T boundary are real. Thus I conclude that although Johnson's estimate of nearly 80% "species" extinction for his megaflora may be somewhat high, it is probably not far off the mark.

There is another important distinction between the vertebrate record and that for megaflora. As I have noted a number of times, the vertebrate record is not of sufficient quality to determine how gradual or abrupt the disappearances were; we can only determine the magnitude and degree of selectivity. This limitation may not apply to plants. The explanation is a bit involved.

To begin, I note that Johnson interpreted the 79% disappearance of his megaflora as an abrupt and catastrophic extinction at the K/T boundary, which he feels is most commensurate with the impact theory. He may be correct, but first three issues must be addressed: (1) There are obvious environmental changes between the K/T boundary megafloras; what do these mean? (2) There is no clear way to determine how much time is represented in the K/T terrestrial boundary clays and coals; does this hinder a diagnosis of gradual or abrupt? (3) confidence in assessments of biotic turnover must be tempered by uncertainties owing to incomplete sections.

Kirk Johnson did address the first of these three issues—that of environmental changes recorded in the rocks at the boundary between latest Cretaceous and earliest Tertiary floras. He and a coauthor noted, "No apparent facies change occurs at the HC II–HC III boundary. The Hell Creek–Fort Union contact [almost at the K/T boundary], however, is defined by a facies change from the poorly drained soils and the meandering river channels of the Hell Creek to swamps, ponds, and meandering to straight distributary river channels of the Ludlow Member [of the Fort Union Formation]" (Johnson and Hickey 1990:438). They went on to say that some of the floral change in the study area was caused by shifting paleoenvironments. They argued, however, that much of the change would thus be independent of facies. I am very concerned about dismissing the facies and the environmental changes seen across the K/T boundary while uncritically accepting the changes as caused by an impact. How can we be sure whether impact or environmental change was the major cause of floral change?

As to the second issue, we simply do not know the interval of time represented by the K/T boundary clays and coals in terrestrial sections and whether there is any substantial gap in the record caused by erosion or nondeposition. All researchers with whom I am familiar agree that this time interval is less than one million years; probably most (me included) would also agree that it is less than 100,000 years. But below this chronological threshold debate begins.

A 1993 paper by the geochronologist Carl Swisher and his colleagues Lowell Dingus and Bob Butler may provide an important clue. They published some high-precision $^{40}Ar/^{39}Ar$ dates for the K/T boundary and for the overlying Paleocene sediments from my study area in eastern Montana. The dates were determined by comparing relative abundances of isotopes of argon (Ar) found in single crystals of the mineral sanidine. The sanidine came from volcanically derived layers found within coals. Unlike other dating techniques that yield results that can be known within a margin of error of only about one million years, these new dates have a margin of error of only tens of thousands of years. Ten thousand years is a long time in human chronology, but it is excellent precision (repeatability) for dates measured in tens of millions of years. Yet we still do not know just how accurate, or close to the K/T boundary, these dates are. Being off only a matter of a few tens of thousands of years makes all the difference if one is arguing (as with the impact theory) that events occurred in months or years.

Finally, as to the third issue, we cannot be certain whether or not the geological sections and the megafloral sampling taking by Kirk Johnson were nearly complete. Interpretations of change must be tempered accordingly. Johnson argued that the "two most significant changes" in his Marmarth sections are between HC III and FU I (K/T boundary), with 21% survival, and between HC II and HC III, with 25% survival. These percentage survivals are cumulative for each of the zones. He felt that "the HC II–HC III survival level is low" because of "continuous turnover throughout HC II, whereas the K/T boundary transition appears more abrupt." I recently asked Johnson to further clarify aspects of these megafloral transitions. He indicated that the HC II–HC III transition moves from a depauperate to a species-rich megaflora, and thus the difference is "reflected in the establishment of a diverse HC III flora rather than extinction of the species in the poorer HC II flora." Johnson may well be correct that the HC II–HC III megafloral survival is low because of continuous turnover during the HC II zone, while the low survival levels at the HC III–FU I (and K/T boundary) is caused by abrupt change. I remain concerned about unconscious biases, however. These can perhaps be seen in one of Johnson's figures.

Figure 9.1 is a representation of figure 4 in Johnson and Hickey's 1990 paper. This representation preserves the content, though not the format, of the original. There appears to be a stepwise change going up section (left to right). Are these steps real? Megafloras are not subject to reworking (the Zombie Effect) because leaf remains are so fragile that they would be reduced to fossil mulch when reworked. The Signor-Lipps Effect can be a factor, however, because stratigraphically highest leaves may not represent the true level of floral extinction; thus, the steps. The way in which the "species" were grouped together may be another reason to suspect that the steps are artifactual, not real.

In their paper, Johnson and Hickey discussed a number of the gaps and floral changes, which I have indicated in figure 9.1. Two changes that the authors themselves regarded as artifactual were at levels -26.25 ft and 164 ft. As can be seen in figure 9.1, there appear to be marked changes at these two levels. The authors attributed these two changes to differences in sample size. Below the marked change, sample sizes were large and above, they were small. Johnson recently told me that the -26.25 ft gap has disappeared through further collecting, while the 164 ft gap remains because no further collections have been made above this level. In their 1990 paper Johnson and Hickey also noted a 10.66 ft gap between the highest Cretaceous HC III zone and the overlying Paleocene FU I zone (figure 9.1). Palynomorphs associated with the HC III megaflora, however, are found higher at the K/T boundary associated with an iridium anomaly. Thus, in a probable example of the Signor-Lipps Effect, one can argue, as Johnson did, that the HC III megaflora continued right up to the K/T boundary.

Combining these factors of sampling artifact and the Signor-Lipps Effect, one can argue that change is significantly more abrupt at the K/T boundary than anywhere else in the entire section. There are, however, other apparent levels with change in this section, which I have noted with question marks in figure 9.1. What are we to make of these? Johnson considered them indicative of continuous change within zone HC II. To me, this seems largely a judgment call. And judgment calls can be perilous when one is up against a geological section that has profound implications for a matter of overwhelming interest and passion in the earth and life sciences. Might those among us who subscribe to the impact theory see abrupt extinction where others of us might well see continuous change?

Geological sections are now being studied in China (Stets et al. 1995) and elsewhere that appear to have a consistent environmental setting across the K/T boundary (see chapter 2). Floral extinctions do occur within these sections, but they tend to be spread out surrounding the boundary. If we accept the Stets group conclusion of gradual or stepwise change across the K/T boundary in China, and also Johnson's argument of abrupt change across the same boundary in North America, then we must conclude that there is no single global pattern of megafloral change at the K/T boundary.

Palynoflora

While the record of megafloral extinction might be as high as 80% in the Western Interior of North America, the palynomorph record suggests only 30%. Because plants bear the pollen and spores, something is obviously amiss here. Although there may be several reasons for this disparity in extinction record, of primary importance is that leaves tend to be structurally more complex than palynomorphs and thus are more easily assignable to lower taxa such as species (Johnson and Hickey 1990).

What palynomorphs lack in power of taxonomic refinement is made up for by their widespread occurrence within the rock record. In fact, so ubiquitous are palynomorphs that many, including me, accept palynology as the best paleontological tool for placing the K/T boundary in any given geological section. In many of these sections in the Western Interior, a palynologically defined K/T boundary is very closely associated with an iridium anomaly (Nichols and Fleming 1990).

Doug Nichols, a palynologist with the U.S. Geological Survey, has been examining the K/T boundary for some time. Along with Farley Fleming, he published an analysis of what is known of palynological changes at the K/T boundary in the Western Interior (Nichols and Fleming 1990). They limited their analysis to sections that showed both a palynological extinction horizon and an iridium anomaly. As these authors note, this combination has been found throughout the Western Interior.

The extinction horizon sometimes coincides with a lithological change, suggesting environmental change. At other places the horizon occurs within the same lithology, for example, a mudstone sequence. Nichols and Fleming convincingly argue that palynological change at the K/T boundary is not controlled by changes in lithology. Their conclusion lends credence to arguments, such as by Kirk Johnson, that megafloral changes at the K/T boundary are not greatly controlled by lithological changes.

Nichols and Fleming surveyed the distribution of more than two hundred taxa (genera and species) of palynomorphs from the palynomorph zone nearest the K/T boundary throughout a nearly 1,500 mile latitudinal gradient from northeastern New Mexico to southern Alberta. They found that the fossil record of plant communities varies along this paleolatitudinal gradient to such an extent that only 25% of the taxa are common to both New Mexico and Alberta. At the K/T boundary they found that plant communities were affected along the entire latitudinal gradient.

Nichols and Fleming conclude that the palynomorph data indicate an abrupt change commensurate with the dust cloud and freezing hypothesized in the original 1980 Alvarez et al. paper. They contend that if the agent of change had been a gradual and progressive deterioration in climate, the consequences should appear first in the harder hit northern plant communities. As it turns out, their prediction may well bear fruit—to the detriment of their theory. A progressive deterioration may indeed be evident in the rock record of the north. A Canadian palynologist, Art Sweet, and colleagues found that from central Alberta south, the palynological change at the K/T boundary is abrupt, as Nichols and Fleming found in the United States. In more northern localities, however, they detected a more gradual loss of species beginning earlier, over the last three to forty vertical feet of Cretaceous rocks (Sweet et al. 1993). As I will discuss in a few pages, floristic changes at high latitudes in Siberia also suggest climatic cooling began earlier at higher latitudes.

The Sweet team calculated that levels of extinction range from 15% to 30% locally, with an aggregate level of 45% for the Western Interior. They further noted, as did Kirk Johnson, that these figures represent minimum extinction levels. Unlike Kirk Johnson's leaves, for palynomorphs we are probably lumping together several species of plants under one species name. As with the megaflora, palynomorph extinctions were not uniform across major plant groups. Extinctions among angiosperms were highest, at 51%; next were gymnosperms, at 36%; least affected were the pteridophytes, with only 25% extinction.

Angiosperms are a monophyletic group including all plants that, among other things, have flowers. Thus we have some sense that a 51% extinction means something biologically. This is not the case for the other two groups, the gymnosperms (which include the conifers) and the pteridophytes (which include the ferns). These latter two groups are not monophyletic, and thus it is not clear what if any biological significance to attach to their levels of extinction. Are the conifers and the pteridophytes circumscribed solely by primitive characters they retain (such as the retention of cold-bloodedness and scales in modern Reptilia, exclusive of birds)? Are they simply convenient aggregations of species that form arbitrary assemblages because they do not fit in other groups? Or are they grouped because of superficial similarities caused by ecological convergence?

Nichols and Fleming (1990) note that the level of extinction among terrestrial plants may or may not constitute a "mass extinction," depending upon how this is defined. Further, whether it is mass extinction or not, the terrestrial plant communities show severe disruption at the K/T boundary. Of all the data from the terrestrial realm, the record of plants in the Western Interior seems to me to present the strongest case that extinction was rapid, not gradual, for the species so affected. It would be logically and scientifically invalid, however, to simply extend this conclusion of rapid extinction to all species in the terrestrial realm—especially if the animal record is not clear one way or the other (as I maintain it is not). As I will discuss in chapter 10, the geological evidence of a dramatic marine regression is strong circumstantial evidence that the biota was under stress long before a presumed impact caused the rapid decline of plant species in the Western Interior.

Nichols and Fleming also attend, in their 1990 analysis, to the fact that many plants have dormancy mechanisms—hardy seeds or spores that might be able to survive in an inactive state until the environmental disruption subsided. Also some plants can die back to the ground and later resprout when conditions are once again favorable. As a paleobotanist colleague once pointed out to me with some glee, you can cut many plants off at their base and they will regrow. If you cut a *Triceratops* in two, the same does not occur.

A final interesting point about the K/T boundary palynological record in the Western Interior is the so-called fern spike, the superabundance (hence, anomaly) of fern spores. In the lowermost Paleocene portions of K/T boundary sections in New Mexico, and somewhat higher up the column in the earliest Paleocene in eastern Montana, R. H. Tschudy and colleagues (Orth et al. 1981; Tschudy et al. 1984) reported palynological assemblages dominated by spores of ferns. The Tschudy team attributed this to rapid recolonization of devastated landscapes after an impact. The fern spike is a regional phenomenon that reaches as far north as southern Canada, but it seems to stop there—although Kirk Johnson has indicated to me that a fern spike is known "further north now." If the fern spike does diminish or disappear at some point further north, however, the pattern could be consistent with an impact in the Yucatán, with effects diminishing the further one moves away from ground zero (Nichols 1991). Early colonization by spore-producing plants, whether after volcanic eruption or road construction, is much in evidence today. This phenomenon is not in itself a unique pattern, other than a general indication of disturbance. Nichols and Fleming (1990) reported that spore abundances range from 70% to 100% in samples found from one to six inches above the palynologically defined extinction event. Such patterns indicate, if anything, general environmental stresses for a measurable, but uncertain length of time after the K/T boundary, but not necessarily devastation from a single impact.

THE MARINE REALM

Apart from controversies over the specifics of dinosaur extinction, the greatest debates are over what happened to marine organisms at the K/T boundary. An obvious and useful first step would be to examine turnover in a phylogenetic context, just as I have done for terrestrial vertebrates. With rare exceptions, however, phylogenetic analyses have not been undertaken for K/T extinctions in the marine realm. More typically the marine record is reviewed by size of the organism, somewhat analogous to palynofloras (or microfloras) versus megafloras. Another set of analyses compare where the organisms live: Are they benthic (bottom dwellers) or do they live in the water column as active swimmers or as planktonic (floating) organisms? The problem with this locational approach is that there are no hard and fast divisions, because some species are planktonic in their larval stages but are benthic as adults. Also as they grow, some species go from the micro to the macro realm in size.

Marine Macrofossils

Macrofossils are the remains of sea organisms that one can usually see without the aid of a microscope. Most groups that leave macrofossils in the fossil record are benthic, and many are outright attached to the bottom. The dominant taxonomic groups of benthic macrofauna recorded near the K/T, all of which are still alive, include mollusks, arthropods, echinoderms, bryozoans, corals, and brachiopods. I left out some important groups (including marine vertebrates) that either do not have a very good fossil record or have not been well studied. Also, some groups, such as the molluscan subgroups of gastropods and bivalves, have important freshwater lineages, but I restrict my examination here to the marine lineages.

The three major groups of mollusks for which turnover rates at the K/T boundary have been calculated are ammonites, bivalves, and gastropods. Percentages of species-level extinctions unfortunately have not been estimated for these major groups, but Jack Sepkoski (1990, figure 3) graphically represented percent extinction at the next more inclusive level—the genus. As with the generic-level comparisons I discussed for Late Cretaceous and early Tertiary vertebrates, caution is required when using higher categories (genus, family, and so on) as they may have no biological meaning. Thus we must keep in mind that these categories are merely proxies for species in biotic analyses.

For ammonite genera Sepkoski shows 100% extinction, for bivalve genera about 55% extinction, and about 35% for gastropod genera. Of these groups, ammonites and bivalves have featured more prominently in discussion of the K/T boundary. Although much of these extinctions may, in fact, be concentrated near the boundary, Sepkoski did not limit his analysis to the boundary itself. His figures apply to last occurrences recorded anywhere within the last stage of the Cretaceous, the Maastrichtian, which may be up to nine million years long.

Ammonites, owing to their shells, are some of the most widely known extinct members of Cephalopoda, the group that also includes the extant octopus, squid, cuttlefish, and the chambered nautilus. The ammonites, which disappeared in the same general time frame as the dinosaurs, showed remarkable diversity in shell form throughout their history. Forms include tightly coiled, loosely coiled, and irregular coiling; some ammonite shells where even straight.

As with dinosaurs in the Western Interior, ammonites on a global scale were declining during the last ten or so million years of the Cretaceous. But did this decline lead to extinction before the K/T boundary? In a 1990 review encompassing four widely separated regions (Europe, Texas, Chile, and Antarctica) an expert on ammonites and the modern chambered nautilus, Peter Ward, concluded that twenty-two species of ammonites are found within about ten feet or less of the K/T boundary—in some cases, within a matter of inches. In one of these areas, Antarctica, it was not even clear whether ammonites became extinct at the K/T boundary. Ward noted in the abstract to his 1990 paper, "The exact level of ammonite extinction, however, cannot be determined; in no section can ammonites be demonstrated to disappear synchronously with the rapid extinction of planktonic foraminiferans and coccoliths occurring at the end of the Cretaceous." (Planktonic foraminiferans and coccolithophorids are some of the microfossils I will discuss in the next section.) Ward also noted that he had some reason to believe that the uppermost Cretaceous rocks were missing. The pattern of disappearance of ammonites before the K/T boundary may of course be real, but what we may also be facing is a sampling problem—the Signor-Lipps Effect—so that ammonites actually did make it right to the boundary. On the other hand, even for those ammonites nearest the boundary, we must be cautious of the Zombie Effect; these fossils may simply have been reworked from older rocks. What Ward has shown is that ammonites retained some diversity to very near, if not demonstrably right to, the K/T boundary.

Another group of mollusks, the bivalves, which includes modern clams, oysters, and scallops, was an important part of the Cretaceous benthic community. David Raup and David Jablonski (1993) studied whether any sort of geographic pattern, especially a latitudinal gradient, is evident in what they termed "end-Cretaceous marine bivalve extinctions."

Raup and Jablonski started by examining the geographic ranges of 340 genera of bivalves from 106 fossil assemblages. They found that extinction levels were highest in equatorial areas of the New World, including the areas nearest the Chicxulub crater. This finding seemed to support the idea of a latitudinal gradient of extinctions at the K/T boundary (which I shall address in the next section). They cautioned, however, that before this apparent geographic extinction gradient is taken as real, we must exclude a group of very specialized bivalves (known as rudists) from the data set. The rudists were important coral-like filter-feeding animals that apparently began to decline several million years before the end of the Cretaceous. They were almost, if not entirely, extinct before the boundary. After excluding the forty-six rudist genera, Raup and Jablonski found virtually no extinction gradient, because the rudists were almost exclusively lower latitude animals.

These authors may well be justified in saying that there is 52–53% bivalve extinction throughout all latitudes at the K/T boundary, but there are some troubling aspects to their study. As they explicitly state, they are interested in extinctions at the end of the Cretaceous, yet they point out that because of sampling problems, they use the record of bivalves for the entire last stage of the Cretaceous, the Maastrichtian. This stage spans an interval of up to nine million years. Thus these authors treat all bivalve extinctions over as much as nine million years as if they occurred at the K/T boundary. Further, these authors used genera rather than species. As I have indicated several times, we must be wary of extinction studies that use higher taxa as surrogates when we do not know how many species are involved.

Some studies that do use species-level data indicate bivalve extinctions occurring prior to the K/T boundary. In the marine fauna from Seymour Island off Antarctica, for example, a number of bivalves disappear well before the boundary (this study will be discussed later in this chapter). One could, of course, invoke the Signor-Lipps Effect and say that such early disappearances owe to sampling error—that all the Seymour Island bivalves really did make it to the boundary. The fauna is of a good enough quality, however, that this possibility is remote. For the sake of argument, let us say that sampling is a factor. If so, the claim should be equally invoked for the rudists—in which case the Signor-Lipps Effect would be taken as the reason for their pre-K/T boundary disappearance, too. It seems inconsistent to exclude one group (the rudists) when that exclusion strengthens the claim of latitudinal gradients, while at the same time lumping together the remaining bivalves for a time interval stretching almost nine million years. This selective exclusion of taxa by David Raup and David Jablonski (1993) is akin to the exclusion made by Sheehan and Fastovsky (1992). Recall (in chapter 8) that the Sheehan and Fastovsky exclusion resulted in an inflation of the extinction differential between freshwater and land-based vertebrates.

The remaining macrofossil groups have not recently received as much close scrutiny as have the mollusks, although some relatively recent studies have examined their basic diversity patterns (Bambach 1985) and extinction patterns (Sepkoski 1990) throughout the Phanerozoic. Jack Sepkoski (1990) shows that all of these other groups also show marked levels of extinction at the end of the Cretaceous. Again, however, it is crucial to recognize that there is a big difference between data calculated for the K/T boundary and data calculated for the last interval of the Cretaceous. Here, too, the calculations represent extinction through the entire last stage of the Cretaceous, about nine million years. As with the Raup and Jablonski study of mollusks, genera were used by Sepkoski as surrogates for species. Thus in this case, once again, the results must be viewed with skepticism because genera may or may not have biological meaning, depending on whether they are or are not monophyletic.

The levels of generic extinctions that I estimated from Sepkoski's figures 2 and 3 in his 1990 paper are 25% for arthropods, 30% for echinoderms, 35% for bryozoans, 35% for corals, and 60% for brachiopods. Arthropods include such familiar sea creatures as lobsters, crabs, shrimp, and barnacles, along with the lesser-known but important ostracods. Of all these macroinvertebrates, arthropods are the least well represented because their body covering, though hard, is more decomposable and more fragile than are the hard parts secreted by the other major taxa. Echinoderms are the starfish, sand dollars, sea urchins, and crinoids. Bryozoans are small, colonial animals that produce encrusting layers on rocks or on other sea creatures; sometimes, however, they do form more erect structures such as fan shapes. Corals are shelly members of the jellyfish group; the most important corals in extinction tabulations are the reef-forming kinds. Finally, the brachiopods are filter-feeding creatures with two opposing shells that superficially resemble the bivalves. The species diversity of brachiopods had already decreased considerably by the Cretaceous, compared to what it had been in the Paleozoic. Nevertheless, the brachiopod group does show some of the higher levels of extinction at the K/T boundary.

Marine Microfossils

Microfossils include a considerable diversity of usually microscopic, single-celled (or sometimes colonial) organisms, or parts of these organisms. Extant organisms that today construct the shelly edifices likely to become tomorrow's fossils are unfamiliar to most people, yet they are a very important part of the marine biota. Also, because the hard parts of their minute remains are a significant part of the material raining out of marine waters onto the seafloor, their fossils are important in tracking past climates. Among the best studied are dinoflagellates, coccolithophorids, diatoms, radiolarians, and foraminiferans (Sleigh 1989; Margulis and Schwartz 1988). The first three groups are algae that can photosynthesize to produce their food and energy. The latter two groups lack chlorophyll and ingest food (obviously minute particles), but often form symbiotic relationships with photosynthesizing species. Silica (glass) forms the skeletons of radiolarians and the cell walls of diatoms, while calcium carbonate forms hard parts in coccolithophorids and some foraminiferans. Accumulations of the carbonate secretors produced such massive chalk outcrops as the White Cliffs of Dover.

The microfossil group that has been most discussed and argued about at the K/T boundary is the foraminiferans. Forams are usually divided into ecological groupings based upon whether they are bottom-dwelling (benthic) species or planktonic species that drift higher in the water column. All forams are single-celled creatures that produce hard tests, less properly referred to as shells. Forams can produce a tremendous rain of microscopic particles onto the seafloor. According to Jack Sepkoski's 1990 compilations, benthic and planktonic forams together suffered about 35% generic extinction at the K/T boundary.

Although there is some disagreement about the level of extinction, the most hotly contested issues about forams involve questions of completeness of their fossil record, reworking (the Zombie Effect), and whether extinction is abrupt or gradual leading up to the K/T boundary. These issues, by now, ought to sound familiar. Whether it is the very spotty record of the giant dinosaurs or the much fuller record of the minuscule forams, questions of completeness of the record, rapidity of extinction, and reworking of fossils continue to plague any analysis of end-Cretaceous events.

The strongest advocates of impact-induced extinctions usually argue that K/T sections are complete and that most, but not all, latest Cretaceous forams that show up in the lower parts of early Tertiary sections are reworked. Scientists who tend to downplay the biological significance of impact-induced foram extinctions contend that there are gaps in the rock and fossil record and that many of the latest Cretaceous forams found in earliest Tertiary rocks are not reworked.

The view that marine K/T boundary sections are often incomplete has been most recently and strongly advocated by Norm MacLeod of the British Museum of Natural History and by Gerta Keller of Princeton University (MacLeod and Keller 1991a,b). Figure 9.2 is a simplified version of a figure that appeared in a *Geology* paper. MacLeod and Keller analyzed the completeness of sediment deposition at and just following the K/T boundary (indicated by the ~65 mya date in figure 9.2) from two major areas of ocean, the continental shelf and the deep sea. The more nearly complete the sediment record, in general, the more nearly complete is the fossil foram record. Continental shelf is a sometimes wide, but always relatively shallow, expanse of submerged continental rock

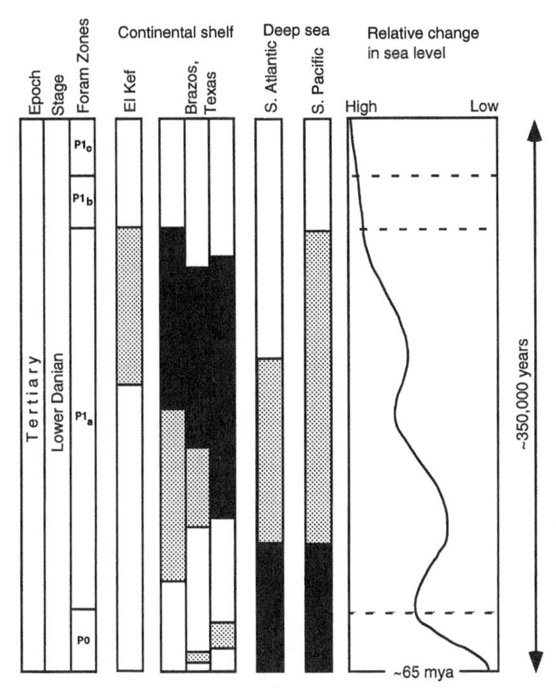

FIGURE 9.2 Incompleteness of deep-sea and continental-shelf records at and beyond the K/T boundary. White portions of columns are most complete and thus have the best forams; black areas are the least complete and thus have the worst records of fossil forams; gray areas are intermediate in completeness and preservation. At and just following the K/T boundary (here shown as ~65 mya), continental shelves and nearby shelf areas offer the better record of fossil foraminiferans because most deep sea cores show gaps where the lower foram zones should be. Thus deep sea cores that are used to argue for abrupt biotic change at the K/T boundary actually record a break in sedimentation and the fossil record—not an abrupt change in biota. After MacLeod and Keller 1991a.

that lies off some coasts—notably, eastern North America—and that separates the continental rock from the true ocean floor of the abyssal depths. Off the western coast of California, by comparison, there is very little if any continental shelf before great depths are reached.

Different sorts of fossils are accumulated on continental shelves versus deep sea regions. Conveniently, El Kef in Tunisia and the Brazos River in Texas (figure 9.2) contain surface exposures on land of what were continental shelf regions sixty-five million years ago. The deep sea sections at the end of the Cretaceous are still deep sea, however, and thus must be obtained by coring.

MacLeod and Keller analyzed several shelf and deep sea sections, some of which are listed in figure 9.2. They concluded that at and just following the K/T boundary the continental shelves and their nearby shelf areas were more complete than were the deep sea sections, and thus better preserved the forams. Places such as El Kef and even Brazos River preserved all or at least parts of the lowest foram zone (P0) in the lowermost Tertiary (white portion of columns in figure 9.2). Most deep sea cores, however, were missing much or all of the lower zones (black portion of columns in figure 9.2). This meant that the deep sea cores that had been used to argue for abrupt biotic change were actually recording only a break in the sediment and fossil record. The shaded areas in the columns are portions of questionable completeness.

The culprit for this discrepancy in quality of preservation between shelf and deep was, once again, sea-level change. Sea level appears to have ebbed until just before the K/T boundary. MacLeod and Keller studied the fossil and rock records close to the latest Cretaceous coastline, which preserve the beginnings of a new sea-level rise in the latest Cretaceous and into the earliest part of the Tertiary. More recently, Keller and Wolfgang Stinnesbeck teamed to refine the timing of sea-level changes near the K/T boundary, based on geological sections in Denmark, the Mediterranean area, Mexico, Brazil, and Antarctica (1995). They found that sea level fell by about three hundred feet some 300,000 years before the end of the Cretaceous and then began to rise again some 100,000 years before the K/T boundary.

You may recall from chapter 8, however, that this rise in sea level does not appear in the geological record in the midcontinent in areas such as Montana and the Dakotas until the early part of the middle Paleocene. This delayed indication of sea-level rise is not surprising, as these inland areas were higher than the sea-level exposures studied by Keller and Stinnesbeck (1995). As the seas began to rise along the outer edges of the continents, the sediment loads from rivers that had formerly made it out to the depths of the ocean were now dumped on the expanded shelf regions. This is also when sharks and their relatives make their reentry into the midcontinent. The lesson from this detective work is that geological sections are not always what they seem.

Both Norm MacLead and Gerta Keller have also been involved in the question of reworking of foraminiferans and other marine fossils at the K/T boundary. This is a stickier issue, and it is one on which I have not yet formed an opinion. As I discussed at length in chapter 3, the issue of reworking in the terrestrial Bug Creek sections in Montana has been a headache to disentangle. We still do not know what is and is not reworked, and thus I have taken the conservative view that until strong evidence appears to the contrary, all distinctively Late Cretaceous elements of the fauna—such as dinosaurs and marsupials—are reworked. I do believe that some Bug Creek specimens are not reworked, but as yet no clear test has been devised to sort out the mess. I am, of course, no expert on the marine K/T boundary, but my experience with reworking of specimens in the terrestrial realm makes me skeptical of marine analyses that do not give this process its due.

Norman MacLeod has not overlooked this possible complication. In a 1994 paper he discussed a number of methods for testing whether forams are reworked. One of the most promising may be the tracking of geochemical signals of stable isotopes, such as isotopes of oxygen, in forams across the boundary. One finds that the relative levels of isotopes of stable elements do change through a geological section, giving a pattern of ratios through time. If a foram is not reworked it should bear the oxygen isotope ratios of the sediment in which it is found and other forams that lived at the same time. If it is reworked, its isotopic ratios should be different from its entombing sediment and the forams that are known to be unreworked. This approach is only beginning, but MacLeod has found that some supposedly reworked forams do bear the isotopic signature of the entombing sediment and contemporaries, suggesting these are survivors of the K/T boundary, not reworked specimens.

At the second of three special conferences on impacts and extinction (called "Snowbird" conferences because the first two were held off-season at the Snowbird ski resort in Utah), a lively discussion erupted between Gerta Keller of Princeton and Jan Smit of the Free University, Amsterdam, about whether planktonic forams disappeared gradually before or abruptly at the K/T boundary in the El Kef section in Morocco. El Kef is important because it has become the marine reference section including the K/T boundary—that is, it has become the standard against which other sections are compared. Such reference sections are important as a starting point for analysis, but this does not mean that they are the best at showing all aspects of the K/T boundary.

Smit argued that most if not all Late Cretaceous forams from El Kef abruptly disappeared at the K/T boundary, while Keller argued that only about one-third of the forams disappear abruptly at the boundary. As this seemed to be an argument over first-level interpretation of data, and thus seemingly testable, a blind test was arranged whereby samples were collected from El Kef—some under the supervision of Smit and some under the supervision of Keller. These samples were then distributed to four workers who were not told from where the samples came in the El Kef section. The blind tests were done under the direction of Bob Ginsburg of the University of Miami. At the third Snowbird conference in February 1994, held in Houston, Bob Ginsburg was unable to personally present the findings of the El Kef tests, so Al Fischer of the University of Southern California read the results. Jan Smit then gave his interpretation of the results, which he regarded as supporting abrupt change. I found the arguments for abrupt change generally compelling, but Keller had already left the meeting to catch a plane, so there was no opportunity to hear an opposing view.

Until the findings of the blind test are published, and until both sides in the debate make their interpretations known, we are left with the unsatisfying result of not knowing the "end of the story." Science is fortunately a reparative process, so there is hope that more definitive conclusions will be forthcoming.

NO MASS EXTINCTION IN THE HIGH LATITUDES

One of the most interesting K/T boundary patterns to emerge in the past five years is restricted to high latitudes, and it is found in both terrestrial and marine realms. Unlike the situation at lower latitudes, there is no *single* horizon in higher latitudes in which one finds major extinctions.

The only problem I see with these high latitude K/T boundary sites, from a selfish perspective, is that there are no vertebrate faunas yet known. Cretaceous vertebrates have, of course, been reported from high latitudes in both northern (Clemens and Nelms 1993) and southern (Vickers-Rich and Rich 1993) hemispheres, but none of the high latitude faunas are at the K/T boundary. The most thoroughly studied high latitude sequences across the K/T boundary occur on Seymour Island. Seymour Island is at the north end of the Antarctic Peninsula, which stretches toward South America. Today, Seymour Island is at 64° south latitude, which is thought to be just about where it was positioned during the waning part of the Cretaceous.

During the 1980s two invertebrate paleontologists, Bill Zinsmeister and Rodney Feldmann, and a host of other earth scientists conducted four paleontological and geological field expeditions to Seymour Island (Zinsmeister and Feldmann 1994). What they found tells a totally different story for the end Cretaceous than the catastrophic extinctions that some argue occurred at lower latitudes. Although there are changes across the K/T boundary in the marine invertebrate faunas and the palynomorphs preserved at Seymour Island, the changes do not appear to be abrupt. Rather, they are spread vertically throughout an almost hundred-foot geological section.

Figure 9.3 is a simplified version from the study made by the antarctic team (Zinsmeister et al. 1989). It shows the pattern of disappearances of species of large invertebrates across the K/T transition zone. There appears to be some stepping of the disappearances. Whether or not the stepping is largely artifactual, there is nevertheless a distinct pattern of continual rather than abrupt extinction. Although I have not reproduced it here, the antarctic team also found a similar pattern of appearances before, throughout, and after the K/T transition zone. There is thus no abrupt change in the marine fossil record of Seymour Island. Further, the pattern of disappearances is not for a single group, but includes a host of invertebrate groups—lobsters, annelid worms, snails, bivalves, ammonites, and nautiloids.

As the Zinsmeister group pointed out, there is no clear place in the Seymour Island sections where one can establish the K/T boundary by using fossils. Their explanation is that other K/T sections are probably condensed and thus indicate extinctions at about the same time even though the extinctions may have been spread out. In contrast, the Seymour Island sections are thick and continuous, providing a temporally more detailed picture of the loss of taxa. One can use an iridium anomaly to identify the K/T boundary in the Seymour Island section, but the fossil organisms do not show a clear indication of mass extinction associated with the anomaly (Elliot et al. 1994). As shown in figure 9.3, the last planktonic foraminiferan that is diagnostic of the Cretaceous, the last in-place ammonite, changes in dinoflagellates ("dinocyst"), and an iridium enrichment all occur at different levels. Ammonites are also found higher in the section, but these are not in place within the rock—that is, they are not demonstrably in their original burial position. Some appear to be reworked because they occur in association with Cretaceous dinoflagellates, but this may not be true for other specimens. The lowest (last diagnostic Cretaceous planktonic foram) to the highest (dinocyst extinction) indicator of the K/T transition is spread over twenty-one feet of stratigraphic section—certainly not a single event.

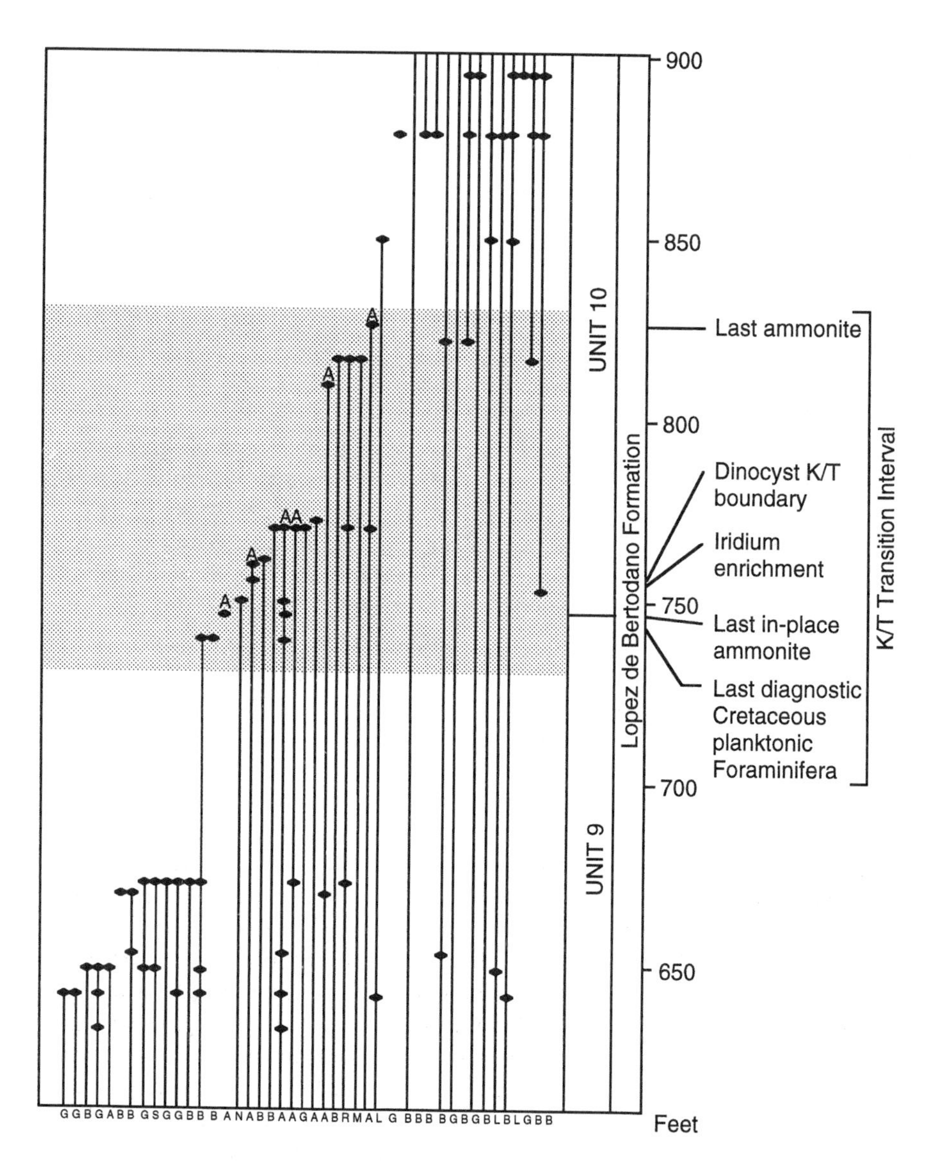

FIGURE 9.3 Marine invertebrate and vertebrate disappearances across the K/T boundary in Antarctica. This simplified plot shows the patterns of disappearances of species of ammonite (A), bivalve (B), gastropod (G), lobster (L), nautiloid (N), scaphopod (M), serpulid (S), and unidentified marine reptile (R) before, across, and after the K/T transition zone on Seymour Island, Antarctica. The black diamonds mark the occurrences of specimens. The vertical lines mark the presumed continuities of existence. It is not clear whether or not stepped disappearances are largely artifactual, but there is a clear pattern of continual extinction. After Zinsmeister et al. 1989.

In their 1989 paper Zinsmeister and colleagues stress the gradual nature of turnover in the Seymour Island sections. The upper graph in figure 9.4 here shows some decrease in the number of taxa across the K/T transition zone, with a resulting lower diversity in the early Tertiary. But this trend is gradual, and it starts well before the hundred-foot vertical block that the authors designate as the K/T transition zone. The lower graph shows a gradual replacement of taxa through the K/T transition. Both first and last occurrences increase throughout the entire section. Last occurrences increase more rapidly as one approaches and passes through the K/T transition zone, however, and this causes the decline in the number of taxa in the upper graph in figure 9.4 over the same interval of time. Once again we see gradual, not abrupt, changes through the K/T boundary, however it is defined.

One could argue that whereas the small iridium spike reported by Elliot et al. (1994) represents the true K/T boundary, all the paleontological data that do not match the spike are artifacts of preservation. Thus the last in-place ammonites and the last diagnostic Cretaceous planktonic foraminiferans occur well below the iridium spike because of inadequacies of sampling associated with the Signor-Lipps Effect; the Cretaceous dinoflagellates and last ammonites occur above the iridium spike because of reworking in accordance with the Zombie Effect. Recently, Marshall (1995) argued in a similar manner in his study of the pattern of ammonite disappearances on Seymour Island. His statistical approach showed that the pattern of ammonite disappearances is consistent with a sudden mass extinction. A computer simulation that he performed on the Seymour Island ammonite fossil record is, however, also consistent with other scenarios, including a gradual pattern of extinction. He further argued that without more intense collecting near the boundary on Seymour Island we cannot distinguish between gradual and sudden extinction.

In a discussion, Charles Marshall clarified this last point. If with intense sampling we greatly increase the density of the samples within the known range of a fossil species, but do not find fossils closer to the K/T boundary, we could eventually statistically eliminate a scenario of sudden mass extinction. If, however, intense sampling also adds more and more ammonites closer and closer to the K/T boundary, a scenario of sudden mass extinction becomes a far greater statistical possibility.

Marshall's approach shows promise for future analyses, but what of our present knowledge of the Seymour Island K/T section? There may be some problems with fossil sampling (e.g., the last ammonites above the K/T boundary), but remember, the Seymour Island section is thick and continuous, and it ranks as one of the richest and most completely

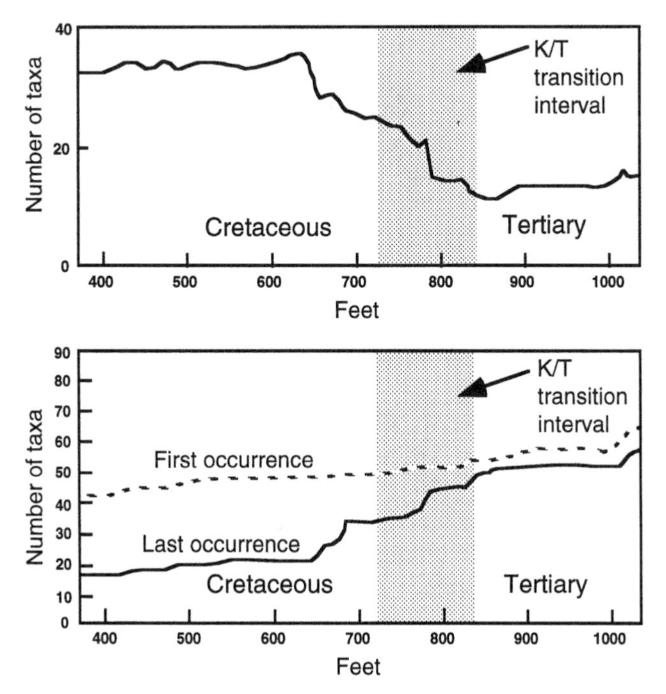

FIGURE 9.4 Evidence of gradual change across K/T boundary in Antarctica. Turnover of invertebrates across the K/T transition zone on Seymour Island, Antarctica, shows gradual rather than abrupt change. *Top:* total number of species. *Bottom:* first and last occurrences. After Zinsmeister et al. 1989.

sampled for macrofossils at the K/T boundary at any latitude. In my view, the most reasonable conclusion based upon present knowledge is that the patterns of disappearances (and appearances) seen on Seymour Island, and probably for higher latitudes in general, indicate a more continual biotic turnover throughout the K/T transition compared to lower latitude assemblages. The completeness of the Seymour Island sections also raises the specter that lower latitude K/T sections are either condensed or incomplete and thus that the abrupt extinction horizons in these areas are more illusory than real.

Gerta Keller, the micropaleontologist involved in the boundary controversy discussed in the previous section, has also studied marine invertebrates surrounding the K/T boundary in the antarctic region (1993a). Her fossil foraminiferans, however, do not come from geological exposures on what is now land on Seymour Island; they come from deep sea drilling cores off the antarctic coast. Her findings echo those of Zinsmeister and colleagues that there is no abrupt mass extinction horizon at the K/T boundary at this high latitude site. Many of the species she studied from before and after the K/T boundary were cosmopolitan. Many seem to have been generalists that had replaced more specialized forms that were perhaps less tolerant of wider ranges in temperature, salinity, and nutrients. Just before the K/T boundary these generalists declined in abundance, with some species becoming dwarfed—an indication of increased environmental stress. There is no evidence of mass mortality or mass extinction among the foraminiferans, however. On the contrary, end-Cretaceous forams continue into the early Tertiary; only then did many finally become extinct. The apparently more specialized species (those with less tolerance of environmental fluctuations) disappear at the K/T boundary at lower latitudes, but the fossil record from higher latitudes clearly shows that this is not a global event.

Large and small marine organisms are not the only fossils recovered from high latitude sites in Antarctica. Pollen and spores of land plants have also been studied. In a terrestrial flora study of Seymour Island, Rosemary Askin and colleagues (1994) found that there is "no evidence of major environmental upheaval." There were, of course, extinctions among land plants, but as with the invertebrates, the turnover was gradual. Moreover, dominant plant taxa remained essentially unchanged through the K/T transition. It is the minor elements of the palynoflora that change gradually through this same sequence (Askin 1990). Sea-level change is implicated as a factor in these floral shifts (e.g., Askin et al. 1994).

The marine realm and terrestrial floras are also represented in K/T sections in New Zealand. The present position of the South Island of New Zealand is today between 42° and 43° south latitude. Like Australia, however, New Zealand drifted north during the past sixty-five million years. Paleogeographic reconstructions place the island almost 20° further south in the Late Cretaceous, essentially at the same high latitudes as Seymour Island (Smith et al. 1981).

Chris Hollis (1993) studied the fate of radiolarians from New Zealand across the K/T boundary. These tiny, one-celled animals grow often intricate outer skeletons (tests) made from glassy silica. Hollis reported that all forty-four species of latest Cretaceous radiolarians make it across the K/T boundary, with most gradually disappearing during the early Paleocene. He speculated that changes in radiolarian populations probably reflect long-term cooling trends related to sea-level change. As with the Seymour Island sections, the fossil evidence drawn from New Zealand radiolarians shows much less extinction at higher latitudes than at lower latitudes across the K/T boundary.

If you recall from the earlier discussion of the palynofloral and megafloral records in the western United States, palynofloral samples tend to underestimate turnover compared to megafloral samples. This is because the simpler morphology of the palynomorphs may be hiding more than one species. Although there is no megaflora reported for Seymour Island, megafloras are now beginning to be studied by Kirk Johnson (1993) in New Zealand. As was the case on Seymour Island, Johnson found minimal palynofloral change across the K/T boundary in New Zealand. He has not yet been able to place a percent extinction level on the megaflora from New Zealand, but he did find that this extinction was, as expected, greater than the palynofloral change, but "substantially less" than that observed for megafloras in the Western Interior of North America.

If we move our investigation to the other, arctic side of the globe, we find that less information has been published about changes at the K/T boundary in the northern high latitudes than in the southern. What is known, however, mirrors the antarctic patterns. Louie Marinkovich (1993) has been studying shallow marine K/T sections ringing the Arctic Ocean. He discovered that a significant number of the bivalves and gastropods that seem to disappear at or before the end of the Cretaceous at lower latitudes survived well into the early Tertiary in the Arctic.

For the Western Hemisphere, much more has been published about the middle and lower latitude fossil floras near the K/T boundary in North America, but there is a considerable literature on Late Cretaceous and Paleogene floras in the high latitudes of Russia. New studies are appearing about these high latitude Russian floras and what they reveal about K/T extinctions. Lena Golovneva recently reported (1994) results of a megafloral study through the latest Cretaceous and early Paleocene, including sampling across the K/T boundary. Eighty-six species were recovered from 124 localities in the high latitude Koryak Upland of northeastern Russia. Although the time frame is not as well established as in North America, Golovneva noted that the age assessment for her floras was based on the occurrence of dinosaurs and correlations to marine rocks containing Late Cretaceous mollusks.

Several of Golovneva's conclusions bear upon K/T boundary extinction issues. First, she found that many of the plants characteristic of the early Paleocene actually appeared before the end of the Cretaceous. These species then increased in abundance during the latest Cretaceous into the early Paleocene. Second, the level of megafloral extinction at the K/T boundary was not greater than at other transitions in the latest Cretaceous in the same region. She found a 31%

megafloral extinction (11 of 36 species) at the K/T boundary; but she also found a 31% extinction (11 of 35 species) and a 48% extinction (16 of 33) for two earlier floral transitions in the latest Cretaceous. Third, she concluded that the "abruptness" of floral change near the K/T boundary was more the result of the appearance of new species rather than the extinction of old ones. My guess is that by "abruptness" she meant the amount of turnover, because her study does not appear to address issues of rates. Fourth, and I believe of most interest, she finds that the change from a latest Cretaceous to early Paleocene flora took place at different times in different regions.

Golovneva indicates that floras typical of the early Paleocene in her study area "preferred cooler conditions." These same species had earlier first appeared at higher latitudes during the latest Cretaceous, spreading southward into middle latitudes as the climate presumably cooled. She concludes that "in middle latitudes the floral change at the K/T boundary was essentially the result of migration rather than evolutionary processes. For that reason the changes of the floristic assemblages in the sections look more abrupt." Although the matter is far from settled, migration (along with an impact) may in part explain the very rapid K/T boundary floral turnover that paleobotanists, such as Kirk Johnson, are seeing at middle latitudes in western North America. This shift toward a cooler adapted flora is recorded first at higher latitudes in the latest Cretaceous, only later does this cool-adapted flora make its way into lower latitudes. Such plants, which were better adapted not only to cooler climates but also to more drastic shifts in light regimes, would have fared much better and spread much more quickly if a photosynthetic crisis occurred as a result of a bolide impact. What little floral evidence that is available regarding climatic change across the K/T boundary in the northern part of the Western Interior does not suggest a shift to a notably cooler flora. According to Kirk Johnson (pers. comm. 1995) all that can be detected is a slight shift from a hot to a warm climate, although as I discussed at length, Johnson does find a dramatic extinction event for the megaflora (about 80%).

Gerta Keller, one of the sharpest critics of the impact scenario and of instantaneous mass extinction in general, has reviewed the differences between high and low latitude survival patterns at the K/T boundary (1993b, 1994). Her findings echo what I discussed in this section—high latitude faunas and floras suggest that the K/T boundary, as recognized by an iridium spike, does not mark a global and abrupt mass extinction. These high latitude finds are showing that, if anything, effects of an impact were far more regionally controlled than is usually portrayed.

X

A Cacophony of Causes

THREE STRIKES AND YOU'RE OUT

Although I do my share of fieldwork and backpacking, I am not a particularly athletic person. Nor do I partake of spectator sports. Thus I am not apt to use a sports metaphor. "Three strikes and you're out," however, seems fitting imagery to express my conclusion about what likely happened at the end of the Cretaceous.

Actually, with all the politically motivated attention to crime, the phrase has taken on another, more foreboding meaning—that of ridding society of career criminals by imprisoning them for life. It is more in this sense that I use the phrase when referring to extinctions at the K/T boundary and, more precisely, to the three major environmental events that may have coincided to cause these extinctions—marine regression, extraterrestrial impact, and massive volcanism. None of these three agents of biological destruction is, in my view, sufficient by itself to be crowned the sole cause of extinctions at the K/T boundary. Were all three necessary, however, for the pattern of extinctions evident at the K/T boundary?

After sorting through the disparate strands of evidence, I have come to conclude that marine regression and an impact are both implicated in the pattern of turnover at the K/T boundary, but I am less certain about volcanism. My uncertainty here is not because I doubt the killing potential of massive volcanism, but because its specific biological effects have not been as closely explored as those of impacts and marine regression. All three events do coincide near the K/T boundary, as figure 10.1 shows. This coincidence does not mean, however, that all three events necessarily occurred simultaneously. A scientifically reliable level of resolution for dating any of these events is at the very best a half million years, which is far less than can be shown in figure 10.1.

In the middle portion of the figure, I show only the last three major mass extinction events from the famous Raup and Sepkoski illustration that compares mass and background extinction (see figure 4.7). It must be remembered, however, that Raup and Sepkoski built their illustration from data pertaining only to families of marine invertebrates. In the lower portion of figure 10.1, I have plotted how much land was added or lost through the last 250 million years as sea level rose and fell. It is based upon the same data plotted in figure 8.1 from the atlas on paleocoastlines by Smith et al. (1994). These authors also provided a tabulation (their table 3) of the geological intervals for their paleocoastline maps. These time intervals varied, which is the reason for the varying widths of the bars in figure 10.1. In their table 3 they also provided an estimate of the total nonmarine area during each interval. I simply determined how much, if any, nonmarine area was lost or gained going from one interval to the next. Figure 10.1 shows very dramatically that the two greatest additions of nonmarine area in the past 250 million years bracket the Late Triassic and Late Cretaceous mass extinctions. The correlation of these events is unmistakable, but causation is another issue. Smith et al. (1994) do not provide data on the extent of nonmarine area back through the Permo-Triassic boundary—thus the question marks in the lower left of figure 10.1. Sea-level curves have been generated for this interval, but there are, as yet, no calculations of changes in nonmarine area. Both Doug Erwin, in his book on the Permo-Triassic extinctions (1993), and Tony Hallam, in his book on Phanerozoic sea-level change (1992), show that the end of the Permian was one of the times of greatest continental exposure in all of the Phanerozoic. There is not a question as to whether there should be a large positive bar in figure 10.1 for the Late Permian, similar to the tall bars that mark the Late Triassic and Late Cretaceous. The uncertainty, rather, pertains only to its height.

The paleocoastline maps used to bracket the interval for the K/T boundary in my composite figure are those shown earlier in figures 8.2 and 8.3. This is an interval of ten million years. As shown in figure 8.3,

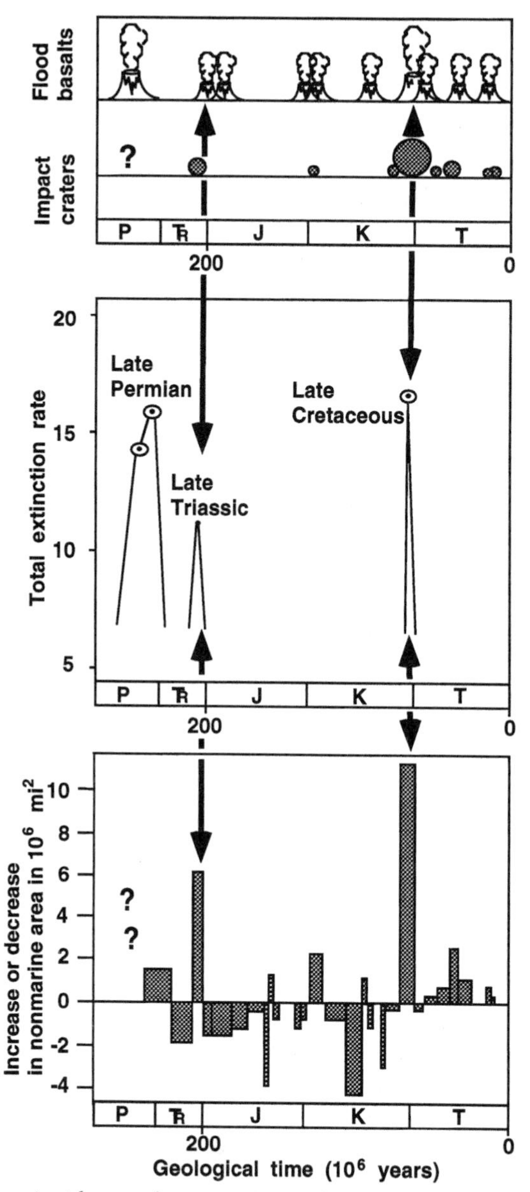

FIGURE 10.1 A coincidence of causes. Coincidence of mass extinction with (1) a tremendous increase in continentality, (2) impact of a very large asteroid or comet, and (3) a marked increase in eruption of flood basalts at or near the K/T boundary. Resolution is at best a half million years; thus the three physical events may have occurred sequentially rather simultaneously at the K/T boundary. Only dramatically increased continentality can currently be related to all three mass extinctions that occurred in the past 250 million years. Heights of volcano plumes and sizes of circles demarcating craters are crudely proportional to the magnitude of these events. The zero mark for nonmarine area is calibrated to no change in continentality. For comparing sizes of other craters and flood basalts, respectively, Chicxulub at the K/T boundary is about 110 miles across and Deccan Traps bracketing the K/T boundary have a volume possibly exceeding 350,000 cubic miles. Source: *Mass extinction patterns after Raup and Sepkoski 1982; rates of increase/decrease of nonmarine area calculated after Smith et al. 1994; impact crater estimates after Grieve and Robertson 1987, Newsom et al. 1990, Raup 1991, Izett et al. 1993, and Hildebrand et al. 1995; flood basalt data after Courtillot 1990 and Campbell et al. 1992.*

sea level during the mid Paleocene (about 60 mya) made a short, but measurable, rise in places such as the midsection of North America. This rise is what brought the sharks back into the Western Interior for a short time. As I noted in chapter 8, sea level reached its nadir near (probably just before) the K/T boundary at about 65 mya. At this time, which is about five million years earlier than the map shown in figure 8.3, continental exposure was even greater. Thus the bar for this interval that I have used in figure 10.1 would be even higher than shown if it depicted just the last five million years of the Cretaceous. Further, the greatest amount of exposure of the lowest, flattest areas would have occurred in only the last tens of thousands of years. The lowest ebbing of the great Late Cretaceous seaways was a geologically very rapid process. If this is hard to envision, recall what occurs as a bathtub empties. As water level begins to recede, the pace is relatively constant on the vertical walls; but as the curved and then flatter areas are reached at the bottom, the last remaining water disappears in a gurgling rush.

At the top of figure 10.1 I have plotted information for known impact craters bigger than about sixteen miles across. I have thus included a few craters that are smaller than the approximate minimum of twenty miles across that David Raup (1991) calculated would cause a 10% extinction. I am not, however, confident that this figure shows absolutely all known impacts of this size and above, as I drew the data from a variety of sources (Grieve and Robertson 1987; Newsom et al. 1990; Raup 1991; Hildebrand et al. 1995; Izett at al. 1993). The relative sizes of the circles correspond to the estimated relative sizes of the craters, with Chicxulub at the K/T boundary being the largest at about 110 miles across (Hildebrand et al. 1995). The figure shows that cratering is not limited to just the Late Triassic and Late Cretaceous mass extinctions, but it is important to recognize that the search for craters has been centered at times of mass extinction, so the record is skewed. In figure 10.1 I left out other evidence of impact because such evidence is even less well known than cratering. For example, although six iridium anomalies have been reported in the Phanerozoic (Raup 1991), only at the K/T boundary is there a correlation among crater, iridium, mass extinction, and shocked quartz. This boundary may thus be unique. Most other iridium spikes in the record are attributable to biological or terrestrial causes (Raup 1991; Wang et al. 1993).

It is clear that the Chicxulub crater (the largest in figure 10.1) is as close to the K/T boundary as the limits of resolution can place it (Swisher et al. 1992). The older Manicouagan crater in Quebec (near the Triassic-Jurassic boundary in figure 10.1) may be related to the Triassic-Jurassic boundary extinctions, but age discrepancies between the crater and extinctions may be as great as ten million years. Finally, there is a big problem with invoking impacts as the sole or even primary cause of mass extinction in general. This is because no impact crater has yet been reported near the Permo-Triassic boundary. The extinction that occurred at this boundary is, however, by everyone's assessment, the granddaddy of all Phanerozoic extinctions. Evidence for an impact (or impacts) may turn up, but for now, its absence calls into question the primacy of impacts in causing all mass extinction. This does not mean that impacts were not part of the history of Earth, only that their primacy in mass extinction is less than has been claimed.

Scientists (I among them) who are unwilling to jump on the impact bandwagon because we take our scientific skepticism seriously are sometimes unfairly portrayed as geological Neanderthals. Peter Ward, in his 1994 book on mass extinction, and Stephen Jay Gould, in his essay in the October 1994 issue of *Natural History* maintain that impacting has not been regarded as a potentially important Earth process, including mass extinction, because establishment geologists have viewed it as somehow violating an important concept in geology known as uniformitarianism. Such an assessment creates the proverbial straw man. Ward and Gould are thus suggesting that most earth scientists retain a nineteenth century view of uniformitarianism, and thus they cannot accept extraterrestrial impacts as the major cause of mass extinction—whatever the evidence. These authors conveniently fail to note that the concept of uniformitarianism has changed since its inception some two hundred years ago.

In one of his best, but probably least read, books, Gould (1987) explores part of the history of this concept. In one of its earlier guises in the last century, the concept of uniformitarianism, as espoused and championed by such greats in geology as James Hutton and Charles Lyell, was that geological processes operating in the past are those still operating today and that these processes were inexorably slow and continuous. When I learned the concept in the late 1960s, it was often reduced to the phrase "the present is the key to the past," but the emphasis was on the continuity of process, not a continuity of rate (slow and gradual). I was able to find my old historical geology text (Spencer 1964); this text explains that in its restricted sense of process, but not including such peripheral issues of rate, uniformitarianism means "that the nature of processes and the physical and chemical laws have remained the same through time."

Stephen Jay Gould and Peter Ward may be partially correct that there is a knee-jerk reaction to catastrophic explanations among earth scientists. They oversimplify the issue, however, when they argue that most objections have been raised against impact-induced mass extinction because it supposedly violates uniformitarianism as the concept is usually understood today. I contend, rather, that the strongest objections came simply from those of us who believe in the value of scientific skepticism and who insist that solid empirical evidence—not the attraction of novelty or excitement—should be the basis for shifting paradigms in science.

Turning to earthly explanations, I used Vincent Courtillot's correlations of flood basalts and mass extinctions (including others than just the big five) from his 1990 review article in the preparation of figure 10.1. This is the basis for the placement of the volcano symbols at the top of this figure. Courtillot did not, however, catalog flood basalts *not* temporally associated with mass extinction, and thus figure 10.1 may imply more correlation than there is. He did not provide a clear sense of the relative size of the various eruptions, but the largest (and hence the larger symbols in figure 10.1) appear to be the Deccan Traps near the K/T boundary on the Indian subcontinent (figure 7.5) and the Siberian Traps near the Permo-Triassic boundary (Campbell et al. 1992).

A RECAP

I must reiterate that our knowledge of vertebrate extinction and survival is limited to the western part of North America—or, as I have called it, the eastern coastline of Laramidia. The records we do have are for terrestrial (including freshwater) communities close to the seaways. The vertebrate extinctions may have reached about 50%. They were highly selective, however, with 75% of the species extinctions concentrated in five of twelve ecologically and evolutionarily dissimilar groups: sharks, lizards, marsupials, and the bird-hipped and reptile-hipped dinosaurs. We have no reliable evidence for the *rate* of vertebrate extinctions at the K/T boundary, by which we might determine whether the die-off was abrupt or gradual. Further, we have absolutely no *global* record of dinosaur extinction. Early reports from such places as China suggest that dinosaurs may actually have survived the K/T boundary outside of North America; but until more detailed work is done and corroborated, the hypothesis of Paleocene dinosaurs remains only speculative.

Among plants, we have a slightly wider record. The megafloral record of leaves suggests very high extinction figures of up to 80% (at the species level) at lower latitudes in North America. The rapidity of the plant extinctions based upon megafloral fossils is still not resolved; but the palynofloral data, at least at lower latitudes, suggests extinctions may have been quite abrupt. As with the vertebrates, preliminary results from China present the possibility that floral extinctions were more gradual there. The results from higher latitudes show more gradual and lower levels of extinction for plants, probably well below 50%.

Within the marine realm, extinctions for many species at the K/T boundary may have been abrupt at lower latitudes, but we must contend with some gaps in the early Tertiary that probably exaggerate the abruptness of extinctions, especially in deep sea cores. As with the plant data, comprehensive high latitude studies show gradual, much reduced levels of extinction through the K/T interval.

These are, as they say, the facts, albeit with a modicum of interpretation about rates of extinction. We still have some way to go in documenting and correlating marine regressions, impacts, and volcanism with many extinction events. Nevertheless, from the kinds of correlations depicted in figure 10.1, combined with the kinds of patterns of extinction and survival we see at the K/T boundary, I believe we can reject outright single-cause explanations for mass extinctions. More than one event correlates with at least the most recent three mass extinctions in Earth history, and all three correlate with the K/T mass extinctions. The extinctions at the K/T boundary thus do not trace back along only one path of causation. Together the sum of the effects broadly quantified in figure 10.1 may explain the extinctions of many plant and animal species. This is especially germane to the K/T boundary, where the documentation is "as good as it gets" for marine regression, impact(s), and increased volcanic activity. Welding the physical and the biological events leading up to the K/T boundary, I can suggest how events may have unfolded.

A SCENARIO OF K/T EXTINCTION

The scenario of K/T extinction I shall present here is slightly modified from one I presented in a contribution to an edited volume (Archibald 1995). I must begin the story well back before the K/T boundary, maybe as much as five million years before, as the midcontinental seaway began to slip away from the interior of North America. Marine regression also occurred on other continents. As the exiting seaways reached the lower-lying, flatter terrain, the rate of exodus quickened; the final stages of withdrawal happened in at most tens of thousands of years.

By this time some vertebrates, especially larger species, were showing obvious stress. As marine regression quickened, the coastal habitats decreased ever more rapidly. The larger vertebrates—the dinosaurs—were the first to experience declines. Our time machine doesn't permit us to know what was happening to vertebrates in more inland areas (there are few or no fossils). The coastal plains dinosaurs certainly were capable of migrating from one shrinking coastal habitat to another, but finally even this could not arrest further decline. Other large vertebrates suffered, too. The Komodo dragon–size lizards and the single exclusively terrestrial turtle, *Basilemys*, also declined in numbers. Smaller terrestrial vertebrates were also declining, but because of shorter life spans and quicker turnover rates, they adapted more quickly to the environmental changes.

As land bridges emerged from the retreating seas, invaders appeared. In North America these were the newly arriving (though still diminutive) archaic ungulates. In the Western Interior at least, the arrival of these ungulates spelled doom for the marsupial lineages that had flourished for some twenty million years. In South America events took a different turn. Both groups of mammals appeared in South America soon after the K/T boundary (replacing the peculiar South American Mesozoic mammals), but here they divided the guilds, with marsupials becoming the carnivores and the ungulates the herbivores. This coevolutionary arrangement lasted for almost fifty million years in South America, with an infusion of only rodents and primates from the outside world until about 5 mya.

Unlike the terrestrial vertebrates, freshwater species faced far less stress during the marine regression, especially because the size of their habitat was increasing dramatically as the epicontinental seas receded. Whole new stream systems developed. Not all aquatic vertebrates fared so well, however. With the loss of close ties to the seas in areas like eastern Montana, sharks and their kin ventured into interior rivers less frequently, as the distance to the sea expanded from tens to thousands of miles.

Plants and nearshore species also were under added stresses as their respective habitats shrank. Certainly some species must have done fine, as new habitats were formed as the seas regressed. We do not, however, have any clear record of these other, more upland, environments. Throughout this time, the waxing and waning of the eruptions of the Deccan Traps introduced further stresses. One such stress was the added particulate matter in the atmosphere, which induced a cooling and drying in some areas of the globe.

Suddenly, a literally earth-shattering event magnified the differences between species doing well and species doing not so well. An asteroid or comet struck the area that today we call the Yucatán. Maybe other pieces of extraterrestrial debris struck elsewhere in short order. Material ejected into the upper atmosphere formed a cover of darkness, screening out the sun to the point that photosynthesis ceased or diminished for many weeks, depending upon location. The effects were especially acute at lower latitudes and closer to the impact, such as in North America. Low-latitude plants, unaccustomed to lower light regimes caused by seasonal changes in sunlight, were especially hard hit. Higher latitude plants, accustomed as they were to seasonally lower light regimes, survived much better—as did the animals that fed upon them. The effects on higher latitude plants and animals was also tempered if the impact happened to have occurred during their season of reduced light—thus winter in either the northern or southern latitudes. Extinction rates for coastal plants in North America soared because of the cumulative effects of continued habitat loss, drying, and (now) loss of sunlight.

With the sudden extinction of many plant species, and the reduction of biomass in an already highly stressed ecosystem, some vertebrate species—most notably, the last large herbivorous nonavian dinosaurs—quickly succumbed. The last nonavian dinosaurs that lived by predation or scavenging followed in short order—with the larger species going first. In some places on the globe the great creatures may have lingered a while longer, but finally, for the first time in more than 150 million years, no large land vertebrates graced the earth. The landscape was open and waiting for evolution's next gambit—mammals.

Epilogue

As I finish writing, my thoughts drift to my impending second visit to the now former Soviet Union. After a brief stop to the favorite city of my previous visit, St. Petersburg (then Leningrad), fellow paleontologist Lev Nessov of St. Petersburg University and I will head to Tashkent, Uzbekistan, in middle Asia. There we are to meet Oleg Tsurak, a herpetologist in Uzbekistan, who will join us for a paleontological adventure.

Our destination, after several days driving from Tashkent, will be Nukus, a town on the Amu Darya, a river just south of the dying Aral Sea. The Aral Sea, once one of the largest lakes in the world, has been shrinking at an alarming rate because of man-made and possibly even natural causes. The countryside is contaminated with pesticides and herbicides from near-futile attempts to grow cotton in an area with the growing season of South Dakota. The purpose of our visit, however, is not to minister to the living but to unearth the long dead from the interior of the Kyzylkum Desert to the east of Nukus.

Several years ago, Lev Nessov found the oldest mammals of the ungulate clade at localities in the Kyzylkum Desert. Their descendants appeared in North America at the close of the Mesozoic, where I believe they wreaked havoc on the hapless native marsupials. Unlike the sixty-five million year old ungulates that appear in North America as the dinosaurs disappear, the older middle Asian species seem to have been living cheek by jowl (or, more accurately, nose to toe) with dinosaurs in Uzbekistan. Lev believes the Uzbek species could be as old as 80–85 million years, which makes them 15–20 million years older than the first ungulates that appeared in North America. Part of our task will be to find more specimens and to collect other fossils and volcanic rock samples for radiometric dating.

Paleontological fieldwork, whether in the United States or in far reaches with romantic names like Kyzylkum Desert, can be hot, dry, even dangerous, and yes, sometimes boring. Nothing cools one's paleontological ardor more quickly than luckless days without fossil finds. As in any other science, the number of failures match or even surpass the triumphs. There are certainly parallels with life; but unlike many endeavors, in science there is always change, building, tearing down, ad infinitum. This is the heart and soul of science.

More than twenty years ago I heard a metaphor at a conference that does some justice to characterizing science in a phrase. The metaphor came from a man whom I regard as this century's greatest paleontologist and one of its premier evolutionists, George Gaylord Simpson. Maybe it stuck with me simply because, as a greenhorn graduate student, I was then in awe of this man. Simpson likened science, unlike most other human pursuits, to an upward spiral. Much of the scientific endeavor entails going back upon itself, as in a spiral, but at the same time there is an overall direction of change. Call it progress if you will or must, but the trajectory does mean we are learning something from our mistakes.

Unfortunately, at this moment, I am not hopeful that the same holds true for the continued existence of a diversity of life on this planet. This may sound doom and gloom, but it need not be if we follow the credo of carpe diem—seize the day. Environmental problems—most of all, overpopulation—need a proactive approach by the very animal that is the agent of such change. Our hope is in accepting that we are part of the system we call life, not a passive observer, or worse, an unassailable overlord.

I might be accused of being a "tree hugger," and possibly I have become one again. More than likely, this was expunged from me in the spring of 1970. As we prepared to celebrate the first Earth Day at my undergraduate institution, Kent State, the politics of war overwhelmed and killed four of us. Such events serve to galvanize—or drown—one's ideals, especially for young people. Everything does sometimes seem to come full circle. I again am deeply concerned with the fate of humanity and that of our planet, but this time it is from the pragmatic view of a paleontologist versed in the ways of extinction. I am no prognosticator of the biological future, but as a paleontologist I can say with the authority of the fossil record that none of us will get out of here alive. The history of life shows in all its incomplete beauty that, whatever the cause, every species eventually changes into something else or it becomes extinct altogether. Either way, each and every newly arisen species eventually disappears. As our knowledge of the fossil record becomes better and better, it is the second fate that seems the likelier for all.

What will be our fate and the fate of the other species that share with us planet Earth? Will we be extinguished by the impact of errant space debris, similar to that which struck Jupiter in July 1994? The answer is "maybe," but there is a troubling side to such speculations. The media and some scientists used this event to build sentiment for increased expenditures on asteroid and comet detection systems, and even to build and aim nuclear-tipped missiles at space. Some of this money may be well spent (I am a fan or space exploration), but there is a new and disturbing view that impacts are the real threat to life on this planet. Extraterrestrial impacts surely did affect Earth's biota significantly in the past and may again in the future, but impacts are only one potential contributor to other equally important physical and biological events occurring on our planet. If we abdicate our biological responsibilities, we run the risk of not realizing that today *we* are unquestionably the major environmental threat to life on this planet—not impacts, not marine regressions, and not volcanic eruptions. It is my fervent hope that we can make a conscious and concerted effort to halt the practices that are hastening the biological collapse of our world. Nothing less will do.

References

Albrecht, G. H. and J. M. A. Miller. Geographic variation in primates: A review with implications for interpreting fossils. In W. H. Kimbel and L. B. Martin, eds., *Species, Species Concepts, and Primate Evolution*, pp. 123–61. New York: Plenum Press.

Alexander, R. M. 1989. *Dynamics of Dinosaurs and Other Extinct Giants*. New York: Columbia University Press.

Alvarez L. W. 1983. Experimental evidence that an asteroid impact led to the extinction of many species 65 million years ago. *Proceedings of the National Academy of Sciences, USA* 80:627–42.

——. 1987. Mass extinctions caused by large bolide impacts. *Physics Today* (July): 24–33.

Alvarez, L. W., W. Alvarez, F. Asaro, and H. Michel. 1980. Extraterrestrial cause for the Cretaceous-Tertiary extinction. *Science* 208:1095–108.

Alvarez, W. 1986. Toward a theory of impact crises. *Eos, Transactions, American Geophysical Union* 67: 649, 653–55, 658.

——. 1991. The gentle art of scientific trepassing. *GSA Today* 1: 29–31, 34.

Alvarez, W. and F. Asaro. 1990. An extraterrestrial impact. *Scientific American* 263:78–84.

Alvarez, W., F. Asaro, P. Claeys, J. M. Grajales-Nishimura., A. Montanari, and J. Smit. 1994. Developments in the KT impact theory since Snowbird II. New developments regarding the KT event and other catastrophes in earth history. *Lunar and Planetary Institute Contribution*, no. 825: 3–5.

Archibald, J. D. 1977. *Fossil Mammalia and Testudines of the Hell Creek Formation, and the Geology of the Tullock and Hell Creek Formations*. Ph.D. diss., University of California, Berkeley.

——. 1981. Earliest known Paleocene mammal site and its implications for noncatastrophic extinctions at the Cretaceous-Tertiary boundary. *Nature* 291:650–52.

——. 1982. A study of Mammalia and geology across the Cretaceous-Tertiary boundary in Garfield County, Montana. *University of California Publications in the Geological Sciences* 122:1–286.

——. 1987. Late Cretaceous (Judithian and Edmontonian) vertebrates and geology of the Williams Fork Formation, N.W. Colorado. In P. M. Currie and E. H. Koster, eds., Fourth Symposium on Mesozoic Terrestrial Ecosystems, Short Papers, *Occasional Papers, Tyrrell Museum of Palaeontology* 3:7–11.

——. 1989a. South American Paleocene dinosaurs, Cretaceous archaic ungulates, or neither. *National Geographic Research* 5:137–38.

——. 1989b. The demise of the dinosaurs and the rise of the mammals. In K. Padian and D. J. Chure, eds., *The Age of Dinosaurs*, Short Courses in Paleontology, Number 2, The Paleontological Society, pp. 162–74.

——. 1993a. The importance of phylogenetic analysis for the assessment of species turnover: A case history of Paleocene mammals in North America. *Paleobiology* 19:1–27.

——. 1993b. Comment on "Major extinctions of land-dwelling vertebrates at the Cretaceous-Tertiary boundary, eastern Montana." *Geology* 21:90–92.

——. 1995. Testing K/T extinction hypotheses using the vertebrate fossil record. In N. MacLeod and G. Keller, eds., *The Cretaceous-Tertiary Mass Extinction: Biotic and Environmental Effects*. New York: W. W. Norton.

Archibald, J. D. and L. Bryant. 1990. Differential Cretaceous-Tertiary extinctions of nonmarine vertebrates: Evidence from northeastern Montana. In V. L. Sharpton and P. Ward, eds., *Global Catastrophes in Earth History: An Interdisciplinary Conference on Impacts, Volcanism, and Mass Mortality*, Special Paper 247, pp. 549–62. Boulder, Colo.: Geological Society of America.

Archibald, J. D. and W. A. Clemens. 1982. Late Cretaceous Extinctions. *American Scientist* 70:377–85.

Archibald, J. D., W. A. Clemens, P. D. Gingerich, D. W. Krause, E. H. Lindsay, and K. D. Rose. 1987. First North American Land Mammal Ages of the Cenozoic Era. In W. O. Woodburne, ed., *Cenozoic Mammals of North America, Geochronology and Biostratigraphy*, pp. 24–76. Los Angeles: University of California Press.

Archibald, J. D. and J. H. Hutchison. 1979. Revision of the genus *Palatobaena* (Baenidae, Testudines), with the description of a new species. *Yale Peabody Museum Postilla* 177:1–19.

Archibald, J. D. and D. L. Lofgren. 1990. Mammalian zonation near the Cretaceous-Tertiary boundary. In T. M. Bown and K. D. Rose, eds., *Dawn of the Age of Mammals in the Northern Part of the Rocky Mountain Interior, North America*, Special Paper 243, pp. 31–50. Boulder, Colo.: Geological Society of America.

Askin, R. A. 1990. Campanian to Paleocene spore and pollen assemblages of Seymour Island, Antarctica. *Review of Palaeobotany and Palynology* 65:105–13.

Askin, R. A., D. H. Elliot, S. R. Jacobson, F. T. Kyte, X. Li, and W. J.

Zinsmeister. 1994. Seymour Island: A southern high-latitude record across the KT boundary. New developments regarding the KT event and other catastrophes in earth history. *Lunar and Planetary Institute Contribution*, no. 825: 7–8.

Bakker, R. T. 1977. Tetrapod mass extinctions—a model of the regulation of speciation rates and immigration by cycles of topographic diversity. In A. Hallam, ed., *Patterns of Evolution as Illustrated by the Fossil Record*, pp. 439–68. Amsterdam: Elsevier.

Bambach, R. K. 1985. Classes and adaptive variety: The ecology of diversification in marine faunas through the Phanerozoic. In J. W. Valentine, ed., *Phanerozoic Diversity Patterns: Profiles in Macroevolution*, pp. 191–253. Princeton: Princeton University Press.

Belt, E. D. 1993. Tectonically induced clastic sediment diversion and the origin of thick, widespread coal beds (Appalachian and Williston basins, USA). In L. E. Frostick and R. J. Steel, eds., *Tectonic Controls and Signatures in Sedimentary Successions*, International Association of Sedimentologists Special Publication 20, pp. 377–97.

Benton, M. J. and J. M. Clark. 1988. Archosaur phylogeny and the relationships of the Crocodylia. In M. J. Benton, ed., *The Phylogeny and Classification of the Tetrapods. Volume 1: Amphibians, Reptiles, Birds*, pp. 295–338. Oxford: Clarendon Press.

Bohor, B. F., E. E. Foord, P. J. Moderski, and D. M. Triplehorn. 1984. Mineralogic evidence for an impact event at the Cretaceous-Tertiary boundary. *Science* 224:867–69.

Bolger, D. T., A. C. Alberts, and M. E. Soulé. 1991. Occurrence patterns of bird species in habitat fragments: Sampling, extinction, and nested species subsets. *American Naturalist* 37:55–166.

Brett-Surman, M. K. and G. S. Paul. 1985. A new family of bird-like dinosaurs linking Laurasia and Gondwanaland. *Journal of Vertebrate Paleontology* 5:133–38.

Brinkman, D. and E. L. Nicholls. 1993. New specimen of *Basilemys praeclara* Hay and its bearing on the relationships of the Nanhsiungchelydidae (Reptila: Testudines). *Journal of Paleontology* 67:1027–31.

Bryant, L. J. 1989. Non-dinosaurian lower vertebrates across the Cretaceous-Tertiary boundary in northeastern Montana. *University of California Publications in Geological Sciences* 134:1–107.

Buffetaut, E. 1994. Comment on "Paleoecological implications of Alaskan terrestrial vertebrate fauna in latest Cretaceous time at high paleolatitudes." *Geology* 22:191.

Campbell, I. H., G. K. Czamanske, V. A. Fedorenko, R. I. Hill, and V. Stepanov. 1992. Synchronism of the Siberian Traps and the Permian-Triassic boundary. *Science* 258:1760–63.

Cappetta, H. 1987. *Handbook of Paleoichthyology, Volume 3B, Chondrichthyes II, Mesozoic and Cenozoic Elasmobranchii*. Stuttgart: Gustav Fischer Verlag.

Carroll, R. L. 1988. *Vertebrate Paleontology and Evolution*. New York: W. H. Freeman.

Carter, N. L., C. B. Officer, C. A. Chesner, and W. I. Rose. 1986. Dynamic deformation of volcanic ejecta from the Toba caldera: Possible relevance to Cretaceous/Tertiary boundary phenomena. *Geology* 14:380–83.

Cherven, V. B. and J. F. Jacob. 1985. Evolution of Paleogene depositional systems, Williston Basin, in response to global sea level changes. In R. M. Flores and S. S. Kaplan, eds., *Cenozoic Paleogeography of the West-Central United States*. Rocky Mountain Paleogeography Symposium, no. 3., pp. 127–70. Rocky Mountain Section of the Society of Economic Paleontologists and Mineralogists.

Chiappe, L. M. 1992. Enantiornithine (Aves) tarsometatarsi and the avian affinities of the Late Cretaceous Avisauridae. *Journal of Vertebrate Paleontology* 12:344–50.

——. 1995. A diversity of early birds. *Natural History* 104:52–55.

Cifelli, R. L. 1990 Cretaceous mammals of southern Utah: I. Marsupials from the Kaiparowits Formation (Judithian). *Journal of Vertebrate Paleontology* 10:295–319.

——. 1993a. Theria of metatherian-eutherian grade and the origin of marsupials. In F. S. Szalay, M. J. Novacek, and M. C. McKenna, eds., *Mammal Phylogeny, Volume 1: Mesozoic Differentiation, Multituberculates, Monotremes, Early Therians, and Marsupials*, pp. 205–15. New York: Springer-Verlag.

——. 1993b. Early Cretaceous mammal from North America and the evolution of marsupial dental characters. *Proceedings of the National Academy* of Science 90:9413–16.

Clemens, W. A. 1964. Fossil mammals of the type Lance Formation, Wyoming: Part I. Introduction and Multituberculata. *University of California Publications in Geological Sciences* 48:1–105.

——. 1966. Fossil mammals of the type Lance Formation, Wyoming: Part II. Marsupialia. *University of California Publications in Geological Sciences* 62:1–122.

——. 1968. A mandible of *Didelphodon vorax* (Marsupialia, Mammalia). *Los Angeles County Museum Contributions in Science*, no. 133: 1–11.

——. 1973. Fossil mammals of the type Lance Formation, Wyoming: Part III. Eutheria and Summary. *University of California Publications in Geological Sciences* 94:1–102.

Clemens, W. A., J. D. Archibald, and L. J. Hickey. 1981. Out with a whimper not a bang. *Paleobiology* 7:293–98.

Clemens, W. A. and Z. Kielan-Jaworowska. 1979. Multituberculata. In J. A. Lillegraven, Z. Kielan-Jaworowska, and W. A. Clemens, eds., *Mesozoic Mammals: The First Two-Thirds of Mammal Evolution*, pp. 99–149. Berkeley: University of California Press.

Clemens, W. A., J. A. Lillegraven, E. H. Lindsay, and G. G. Simpson. 1979. Where, when, and what: A survey of known Mesozoic mammal distribution. In J. A. Lillegraven, Z. Kielan-Jaworowska, and W. A. Clemens, eds., *Mesozoic Mammals: The First Two-Thirds of Mammal Evolution*, pp. 7–58. Berkeley: University of California Press.

Clemens, W. A. and L. G. Nelms. 1993. Paleoecological implications of Alaskan terrestrial vertebrate fauna in latest Cretaceous time at high paleolatitudes. *Geology* 21:503–6.

Closs, G. P. and P. S. Lake. 1993. Spatial and temporal variation in the structure of an intermittent-stream food web. *Ecological Monographs* 64:1–21.

CoBabe, E. A. and D. E. Fastovsky. 1987. *Ugrosaurus olsoni*, a new ceratopsian (Reptilia: Ornithischia) from the Hell Creek Formation of eastern Montana. *Journal of Paleontology* 61:148–54.

Coombs, Jr., W. P. 1978. The families of the ornithischian dinosaur order Ankylosauria. *Palaeontology* 21:143–70.

Cooper, J. F. 1826. *The Last of the Mohicans: A Narrative of 1757*. Philadelphia: H. C. Carey and I. Lea.

Courtillot, V. E. 1990. A volcanic eruption. *Scientific American* 263:85–92.

Cox, G. W. 1993. *Conservation Ecology Biosphere and Biosurvival*. Dubuque, Iowa: Wm. C. Brown.

Currie, P. J. and P. Dodson. 1984. Mass death of a herd of ceratopsian dinosaurs. In W.-E. Reif and F. Westphal, eds., *Third Symposium on Mesozoic Terrestrial Ecosystems*, pp. 61–66. Tübingen: Attempto Verlag Tübingen GmbH.

Cutler, A. H. and A. K. Behrensmeyer. 1994. Bone beds at the boundary: Are they a real expectation? Developments in the KT impact theory since Snowbird II. New developments regarding the KT event and other catastrophes in earth history. *Lunar and Planetary Institute Contribution*, no. 825: 28.

Cvancara, A. M. and J. W. Hoganson. 1993. Vertebrates of the Cannonball Formation (Paleocene) in North and South Dakota. *Journal of Vertebrate Paleontology* 13:1–23.

Darwin, C. R. and A. R. Wallace. 1858. On the tendency of species to form varieties; and on the perpetuation of varieties and species by natural means of selection. *Journal of the Proceedings of the Linnean Society of London, Zoology* 3:45–62.

Desmond, A. and J. Moore. 1991. *The Life of a Tormented Evolutionist: Darwin*. New York: Warner Books.

D'Hondt, S., H. Sigurdsson, A. Hanson, S. Carey, and M. Pilson. 1994a. Sulfate volatilization, surface-water acidification, and extinction at the KT boundary. New developments regarding the KT event and other catastrophes in earth history. *Lunar and Planetary Institute Contribution*, no. 825: 29–30.

D'Hondt, S., M. E. Q. Pilson, H. Sigurdsson, A. K. Hanson, Jr., and S. Carey. 1994b. Surface water acidification and extinction at the Cretaceous-Teriary boundary. *Geology* 22:983–86.

Dodson, P. M. 1991. Maastrichtian dinosaurs. *Geological Society of America Abstracts with Programs* 23 (5): 184–85.

Dodson, P. M. and P. J. Currie. 1990. Neoceratopsia. In D. B. Weishampel, P. Dodson, and H. Osmólska, eds., *The Dinosauria*, pp. 593–618. Berkeley: University of California Press.

Eaton, J. G. 1987. The Campanian-Maastrichtian boundary in the western interior of North America. *Newsletters on Stratigraphy* 18:31–39.

Eaton, J. G. and J. I. Kirkland. 1993. Faunal changes across the Cenomanian-Turonian (Late Cretaceous) boundary, southwestern Utah. *Geological Society of America Abstracts with Programs* 25 (5): 33–34.

Eldredge, N. 1989. *Macroevolutionary Dynamics, Species Niches, and Adaptive Peaks*. New York: McGraw-Hill.

Elliot, D. H., R. A. Askin, F. T. Kyte, and W. J. Zinsmeister. 1994. Iridium and dinocysts at the Cretaceous-Tertiary boundary on Seymour Island, Antarctica: Implications for the K-T event. *Geology* 22:675–78.

Erickson, B. R. 1972. The lepidosaurian reptile *Champsosaurus* in North America. *Monograph of the Science Museum of Minnesota* 1:1–91.

Erwin., D. H. 1993. *The Great Paleozoic Crisis: Life and Death in the Permian*. New York: Columbia University Press.

Estes, R. 1964. Fossil vertebrates from the Late Cretaceous Lance Formation, eastern Wyoming. *University of California Publications in Geological Sciences* 49:1–187.

——. 1969a. Relationships of two Cretaceous lizards (Sauria, Teiidae). *Breviora* 317:1–8.

——. 1969b. A new fossil discoglossid frog from Montana and Wyoming. *Breviora* 328:1–7.

——. 1982. Systematics and paleogeography of some fossil salamanders and frogs. *National Geographic Society Research Reports* 14:191–210.

Estes, R. and P. Berberian. 1970. Paleoecology of a Late Cretaceous community from Montana. *Breviora* 329:1–35.

Estes, R., P. Berberian, and C. Meszoely. 1969. Lower vertebrates from the Late Cretaceous Hell Creek Formation, McCone County, Montana. *Breviora* 337:1–33.

Evans, S. E. 1988. The early history and relationships of the Diapsida. In M. J. Benton, ed., *The Phylogeny and Classification of the Tetrapods, Volume 1: Amphibians, Reptiles, Birds*, pp. 221–60. Oxford: Clarendon Press.

Fastovsky, D. E. and K. McSweeny. 1987. Paleosols spanning the Cretaceous-Paleogene transition, eastern Montana and western North Dakota. *Geological Society of America Bulletin* 99:66–77.

Fastovsky, D. E. and P. M. Sheehan. 1994. Habitat vs. asteroid fragmentation in vertebrate extinctions at the KT boundary: The good, the bad, and the untested. New developments regarding the KT event and other catastrophes in earth history. *Lunar and Planetary Institute Contribution*, no. 825: 36–37.

Feduccia, A. 1995. Explosive evolution in Tertiary birds and mammals. *Science* 267:637–38.

Flessa, K. W. and D. Jablonski. 1983. Extinctions are here to stay. *Paleobiology* 9:315–21.

Forey, P. L. 1977. The osteology of *Notelops* Woodward, *Rhacolepis* Agassiz and *Pachyrhizodus* Dixon (Pisces; Teleostei). *British Museum (Natural History) Bulletin of Geology* 28 (2): 125–204.

Forster, C. A. 1993. Taxonomic validity of the ceratopsid dinosaur *Ugrosaurus olsoni* (CoBabe and Fastovsky). *Journal of Paleontology* 67:316–18.

Fox, R. C. 1989. The Wounded Knee Local Fauna and mammalian evolution near the Cretaceous-Tertiary boundary, Saskatchewan, Canada. *Palaeontographica*, Abt. A., 208:11–59.

Gaffney, E. S. 1972. The systematics of the North American family Baenidae (Reptilia, Cryptodira). *Bulletin of the American Museum of Natural History* 147:245–319.

Gaffney, E. S. and P. A. Meylan. 1988. A phylogeny of turtles. In M. J. Benton, ed., *The Phylogeny and Classification of the Tetrapods, Volume 1: Amphibians, Reptiles, Birds*, pp. 157–219. Oxford: Clarendon Press.

Gauthier, J. 1986. Saurischian monophyly and the origin of birds. In K. Padian, ed., The origin of birds and the evolution of flight, *Memoirs of the California Academy of Sciences*, no. 8: 1–55.

Gauthier, J., R. Estes, and K. de Queiroz. 1988. A phylogenetic analysis of Lepidosauromorpha. In R. Estes and G. Pregill, eds., *Phylogenetic Relationships of the Lizard Families: Essays Commemorating Charles L. Camp*, pp. 15–98. Stanford: Stanford University Press.

Gill, J. R. and W. A. Cobban. 1973. Stratigraphic and geologic history of the Montana Group and equivalent rocks, Montana, Wyoming, and North and South Dakota. *United States Geological Survey Professional Paper* 776:1–37.

Gilmore, C. W. 1910. *Leidyosuchus sternberghii*, a new species of crocodile from the Ceratops beds of Wyoming. *Proceedings of the United States National Museum* 38:485–502.

Glen, W. 1990. What killed the dinosaurs? *American Scientist* 78:354–70.

Glen, W., ed. 1994. *Mass Extinction: How Science Works in a Crisis.* Stanford: Stanford University Press.

Golovneva, L. B. 1994. The flora of the Maastrichtian-Danian deposits of the Koryak Upland, northeast Russia. *Cretaceous Research* 15:89–100.

Gould, S. J. 1987. *Time's Arrow, Times Cycle: Myth and Metaphor in the Discovery of Geological Time.* Cambridge, Mass.: Harvard University Press.

——. 1989. *Wonderful Life: The Burgess Shale and the Nature of History.* New York: W. W. Norton.

——. 1994. Jove's thunderbolts: Comet Shoemaker-Levy 9 was a pat on the back for science. *Natural History* 103:6–12.

Grieve, R. A. F. and P. B. Robertson. 1987. Terrestrial impact structures. *Geological Survey of Canada, Map* 1658A.

Gurnis, M. 1994. Phanerozoic marine inundation of continents driven by dynamic topography above subduction slabs. *Nature* 364:589–93.

Hall, E. R. and K. R. Kelson. 1959. *The Mammals of North America.* New York: Ronald Press.

Hallam, A. 1992. *Phanerozoic Sea-Level Changes.* New York: Columbia University Press.

Hansen, H. J. 1991. Diachronous disappearance of marine and terrestrial biota at the Cretaceous-Tertiary boundary. In Z. Kielan-Jaworowska, N. Heintz, and H. A. Nakrem, eds., Fifth symposium on Mesozoic terrestrial ecosystems and biota (extended abstracts), *Contributions from the Paleontological Museum, University of Oslo*, no. 364: 31–32.

Haq, B. U., J. Hardenbol, and P. R. Vail. 1988. Mesozoic and Cenozoic chronostratigraphy and eustatic cycles. In W. K. Wilgus et al., eds., *Sea-level Changes: An Integrated Approach*, Special Publication 42, pp. 71–108. Tulsa, Okla.: Society of Economic Paleontologists and Mineralogists.

Harland, W. B., R. L. Armstrong, A. V. Cox, L. E. Craig, A. G. Smith, and D. G. Smith. 1989. *A Geologic Time Scale, 1989.* Cambridge: Cambridge University Press.

Hay, O. P. 1908. The fossil turtles of North America. *Carnegie Institution of Washington, Publication* 75:1–568, 113 plates.

Hershkovitz, P. 1977. *Living New World Monkeys (Platyrrhini) with an Introduction to Primates*, vol. 1. Chicago: University of Chicago Press.

Hildebrand, A. R. 1993. The Cretaceous/Tertiary boundary impact (or the dinosaurs didn't have a chance). *Journal of the Royal Astronomical Society of Canada* 87:77–118.

Hildebrand, A. R., M. Pilkington, M. Conners, C. Oritz-Aleman, and R. E. Chavez. 1995. Size and structure of the Chicxulub crater revealed by horizontal gravity gradients and cenotes. *Nature* 376:415–17.

Hoffman, A. 1984. Mass extinction: More publicity than progress? *Zentralblatt für Geologie und Paläontologie, Teil 2: Paläontolgie* 5/6:211–24.

Hollis, C. J. 1993. Radiolarian faunal change through the K-T transition in eastern Marlborough, New Zealand. *Geological Society of America Abstracts with Programs* 25 (6): 295.

Holtz, T. R., Jr. 1994. The phylogenetic position of the Tyrannosauridae: Implications for theropod systematics. *Journal of Paleontology* 68:1100–17.

Horner, J. R. 1992. Cranial morphology of *Prosaurolophus* (Ornithischia: Hadrosauridae) with descriptions of two new hadrosaurid species and an evaluation of hadrosaurid phylogenetic relationships. *Museum of the Rockies Occasional Paper*, no. 2: 1–119.

Horner, J. R. and J. Gorman. 1988. *Digging Dinosaurs*. New York: Workman Publishing.

Horner, J. R. and D. Lessem. 1993. *The Complete T. rex*. New York: Simon and Schuster.

Horner, J. R., D. J. Varricchio, and M. Goodman. 1992. Marine transgressions and the evolution of Cretaceous dinosaurs. *Nature* 358:59–61.

Hurlbert, S. H. and J. D. Archibald. 1995. No statistical support for sudden (or gradual) extinction of dinosaurs. *Geology* 23: in press.

Hutchison, J. R. 1982. Turtle, crocodilian, and champsosaur diversity changes in the Cenozoic of the north-central region of western United States. *Palaeogeography, Palaeoclimatology, Palaeoecology* 37:149–64.

——. 1993. Avisaurus: A "dinosaur" grows wings. *Journal of Vertebrate Paleontology* 13 (supplement to no. 3): 43A.

Hutchison, J. R. and J. D. Archibald. 1986. Diversity of turtles across the Cretaceous/Tertiary boundary in northeastern Montana. *Palaeogeography, Palaeoclimatology, Palaeoecology* 55:1–22.

Ivany, L. C. and R. J. Salawitch. 1993. Carbon isotopic evidence for biomass burning at the K-T boundary. *Geology* 21:487–90.

Izett, G. A., W. A. Cobban, J. D. Obradovich, and M. J. Kunk. The Manson impact crater structure: 40Ar/39Ar age and its distal impact ejecta in the Pierre Shale in southeastern South Dakota. *Science* 262:729–32.

Johnson, K. R. 1992. Leaf-fossil evidence for extensive floral extinction at the Cretaceous-Tertiary boundary, North Dakota. *Cretaceous Research* 13:91–117.

——. 1993. High-latitude deciduous forests and the Cretaceous-Tertiary boundary in New Zealand. *Geological Society of America Abstracts with Programs* 25 (6): 295.

Johnson, K. R. and L. J. Hickey. 1990. Megafloral change across the Cretaceous/Tertiary boundary in the northern Great Plains and Rocky Mountains, USA. In V. L. Sharpton and P. Ward, eds., *Global Catastrophes in Earth History: An Interdisciplinary Conference on Impacts, Volcanism, and Mass Mortality*, Special Paper 247, pp. 433–44. Boulder, Colo.: Geological Society of America.

Keller, G. 1993a. The Cretaceous-Tertiary boundary transition in the Antarctic Ocean and its global implications. *Marine Micropaleontology* 21:1–45.

——. 1993b. K/T boundary mass extinction restricted to low latitudes? *Geological Society of America Abstracts with Programs* 25 (6): 296.

——. 1994. Global biotic effects of the KT boundary event: Mass extinction restricted to low latitudes? New developments regarding the KT event and other catastrophes in earth history. *Lunar and Planetary Institute Contribution*, no. 825: 57–58.

Keller, G. and W. Stinnesbeck. 1995. Sea level changes, clastic deposits, and megatsunamis across the Cretaceous-Tertiary boundary. In N. MacLeod and G. Keller, eds., *The Cretaceous-Tertiary Mass Extinction: Biotic and Environmental Effects*. New York: W. W. Norton.

Kielan-Jaworowska, Z. and D. Dashzeveg. 1989. Eutherian mammals from the Early Cretaceous of Mongolia. *Zoologica Scripta* 18:347–55.

Kielan-Jaworowska, Z. and P. P. Gambaryan. 1994. Postcranial anatomy and habits of Asian multituberculate mammals. *Fossils and Strata* 36:1–92.

Kirkland, J. I. 1987. Upper Jurassic and Cretaceous lungfish tooth plates from the western interior, the last dipnoan faunas of North America. *Hunteria* 2:1–16.

Lambert, D. 1990. *The Dinosaur Data Book*. New York: Avon Books.

Lanham, U. 1973. *The Bone Hunters*. New York: Columbia University Press.

Lécuyer, C., P. Grandjean, J. R. O'Neil, H. Cappetta, and F. Martineau. 1993. Thermal excursions in the ocean at the Cretaceous-Tertiary boundary (northern Morocco): d18O record of phosphatic fish debris. *Palaeogeography, Palaeoclimatology, Palaeoecology* 105:235–43.

Lewis, J. S., G. H. Watkins, H. Hartman, and R. Prinn. 1982. Chemical consequences of major impact events on Earth. In L. T. Silver and P. H. Schultz, eds., *Geological Implications of Impacts of Large Asteroids and Comets on the Earth*, Special Paper 190, pp. 215–21. Boulder, Colo.: Geological Society of America.

Lillegraven, J. A. 1969. Latest Cretaceous mammals of upper part of Edmonton Formation of Alberta, Canada, and review of marsupial-placental dichotomy in mammalian evolution. *University of Kansas Paleontological Contribution*, Art. 50, Vert. 12: 1–122.

Lillegraven, J. A. and M. C. McKenna. 1986. Fossil mammals from the "Mesaverde" Formation (Late Cretaceous, Judithian) of the Bighorn and Wind River basins, Wyoming, with definitions of the Late Cretaceous North American Land-Mammal "Ages." *American Museum Novitates*, no. 2840: 1–68.

Lofgren, D. L. 1995. *The Bug Creek Problem and the Cretaceous-Tertiary Transition at McGuire Creek, Montana*. University of California Publications in the Geological Sciences 140:1–200.

Lofgren, D. L., C. L. Hotton, and A. C. Runkel. 1990. Reworking of Cretaceous dinosaurs into Paleocene channel deposits, upper Hell Creek, Montana. *Geology* 8:874–77.

MacFadden, B. J. 1992. *Fossil Horses Systematics, Paleobiology, and Evolution of the Family Equidae.* Cambridge: Cambridge University Press.

MacLeod, N. 1994. An evaluation of criteria that may be used to identify species surviving a mass extinction. New developments regarding the KT event and other catastrophes in earth history. *Lunar and Planetary Institute Contribution,* no. 825: 75–77.

MacLeod, N. and G. Keller. 1991a. Hiatus distributions and mass extinctions at the Cretaceous/Tertiary boundary. *Geology* 19:497–501.

——. 1991b. How complete are the Cretaceous/Tertiary boundary sections? A chronostratigraphic estimate based on graphic correlation. *Geological Society of America Bulletin* 103:66–77.

Margulis, L. and K. V. Schwartz. 1988. *Five Kingdoms: An Illustrated Guide to the Phyla of Life on Earth.* New York: W. H. Freeman.

Marinovich, Jr., L. 1993. Delayed extinction of Mesozoic marine mollusks in the Paleocene Arctic Ocean Basin. *Geological Society of America Abstracts with Programs* 25 (6): 295.

Marshall, C. R. 1995. Distinguishing between sudden and gradual extinctions in the fossil record: Predicting the position of the Cretaceous-Tertiary iridium anomaly using the ammonite fossil record on Seymour Island, Antarctica. *Geology* 23:731–34.

Marshall, L. G. and C. de Muizon. 1988. The dawn of the Age of Mammals in South America. *National Geographic Research* 4:23–55.

Maryn'ska, T. 1990. Pachycephalosauria. In D. B. Weishampel, P. Dodson, and H. Osmólska, eds., *The Dinosauria,* pp. 564–77. Berkeley: University of California Press.

McGookey, D. P., J. D. Haun, L. A. Hale, H. G. Goodell, D. G. McGubbin, R. J. Weimer, and G. R. Wulf. 1972. Cretaceous system. In *Geologic Atlas of the Rocky Mountain Region,* pp. 190–228. Denver: Rocky Mountain Association of Geologists.

Miller, R. I. 1978. Applying island biogeographic theory to an East African reserve. *Environmental Conservation* 5:191–95.

Molnar, R. E. and K. Carpenter. 1989. The Jordan theropod (Maastrichtian, Montana, USA) referred to the genus *Aublysodon. Geobios* 22:445–54.

Moyle, P. B. and J. J. Cech, Jr. 1988. *Fishes: An Introduction to Ichthyology.* Englewood Cliffs, New Jersey: Prentice Hall.

Neill, W. T. 1958. The occurrence of amphibians and reptiles in saltwater areas, and a bibliography. *Bulletin of Marine Science of the Gulf and Caribbean* 8:1–97.

Nelson, J. S. 1984. *Fishes of the World.* New York: John Wiley and Sons.

Nessov, L. A. 1984. Upper Cretaceous pterosaurs and birds from central Asia. *Translation of* Pterozavry i ptitsy pozdnego mela Sredney Azii. *Paleontologicheskii Zhurnal* 1:47–57.

——. 1987. Results of search and study of Cretaceous and early Paleocene mammals on the territory of the USSR. *Ezhegodnik Vsesoyuznogo Paleontologicheskogo Obshchestva* 30:199–219 (in Russian).

Nessov, L. A. and L. B. Golovneva. 1990. History of the flora, vertebrates, and climate in the Late Senonian of the northeastern Koryak Uplands. In V. A. Krassilov, ed., *Kontinental'ny'i myel SSSR. Sboznik nauchny'kh tzudov* [Continental Cretaceous of the USSR. Collection of Research Papers], pp. 191–212 (in Russian).

Newsom, H. E., G. Graup, D. A. Iseri, J. W. Geissman, and K. Keil. 1990. The formation of the Ries Crater, West Germany; evidence of atmospheric interactions during a large cratering event. In V. L. Sharpton and P. Ward, eds., *Global Catastrophes in Earth History: An Interdisciplinary Conference on Impacts, Volcanism, and Mass Mortality*, Special Paper 247, pp. 195–206. Boulder, Colo.: Geological Society of America.

Nichols, D. J. 1991. Palynology supports catastrophic event at Cretaceous-Tertiary boundary: An overview of ten years of research in the United States. *Geological Society of America Abstracts with Programs* 23 (5): 357.

Nichols, D. J. and R. F. Fleming. 1990. Plant microfossil record of the terminal Cretaceous event in the western United States and Canada. In V. L. Sharpton and P. Ward, eds., *Global Catastrophes in Earth History: An Interdisciplinary Conference on Impacts, Volcanism, and Mass Mortality*, Special Paper 247, pp. 445–55. Boulder, Colo.: Geological Society of America.

Nichols, D. J., L. J. Hickey, L. J. McSweeney, and J. A. Wolfe. 1992. Plants at the K/T boundary: Discussion and reply. *Nature* 356:295–96.

Norell, M. A., J. M. Clark, D. Demberelyin, B. Rhinchen, L. M. Chiappe, A. R. Davidson, M. C. McKenna, P. Altangerel, and M. J. Novacek. 1994. A theropod dinosaur embryo and the affinities of the Flaming Cliffs dinosaur eggs. *Science* 266:779–82.

Norell, M. A., J. M. Clark, and J. H. Hutchinson. 1994. The Late Cretaceous alligatoroid *Brachychampsa montana* (Crocodylia): New material and putative relationships. *American Museum Notitates* 3116:1–26.

Norman, D. 1985. *The Illustrated Encyclopedia of Dinosaurs.* New York: Crescent Banks.

Officer, C. B. and A. A. Ekdale. 1986. Comment on "Cretaceous extinctions: Evidence for wildfires and search for meteoritic material." *Science* 234:262–63.

Orth, C. J., J. S. Gilmore, J. D. Knight, C. L. Pillmore, R. H. Tschudy, and J. E. Fassett. 1981. An iridium anomaly at the palynological Cretaceous-Tertiary boundary in northern New Mexico. *Science* 214:1341–43.

Osborn, H. F. 1905. Tyrannosaurus and other Cretaceous carnivorous dinosaurs. *Bulletin of the American Museum of Natural History* 21:259–65.

Osmólska, H. and R. Barsbold. 1990. Troodontidae. In D. B. Weishampel, P. Dodson, and H. Osmólska, eds., *The Dinosauria*, pp. 259–79. Berkeley: University of California Press.

Ostrom, J. H. 1966. Functional morphology and evolution of the ceratopsian dinosaurs. *Evolution* 20:290–308.

Otte, D. and J. A. Endler, eds. 1989. *Speciation and Its Consequences.* Sunderland, Mass.: Sinauer Associates.

Owen, R. 1841 [1842]. Report on British fossil reptiles. Part II. *Report of the British Association for the Advancement of Science* 11:60–204.

Padian, K. and W. A. Clemens. 1985. Terrestrial vertebrate diversity: Episodes and insights. In J. W. Valentine, ed., *Phanerozoic Diversity Patterns: Profiles in Macroevolution*, pp. 41–96. Princeton: Princeton University Press.

Perlman, D. 1993. Volcanoes may have stolen dinosaurs' oxygen supply. *San Francisco Chronicle*, Final Edition, 28 October 1993, p. A5.

Pillmore, C. L., M. G. Lockley, R. F. Fleming, and K. R. Johnson. 1994. Footprints in the rocks: New evidence from Raton Basin that dinosaurs flourished on land until the terminal Cretaceous impact event. New developments regarding the KT event and other catastrophes in earth history. *Lunar and Planetary Institute Contribution*, no. 825: 90.

Pollman, C. D. and D. E. Canfield, Jr. 1991. Florida. In D. F. Charles, ed., *Acidic Deposition and Aquatic Ecosystems Regional Case Studies*, pp. 367–416. New York: Springer-Verlag.

Prasad, G. V. R., J. J. Jaegar, A. Sahni, E. Gheerbrant, and C. K. Khajuira. 1994. Eutherian mammals from the Upper Cretaceous (Maastrichtian) intertrappean beds of Naskal, Andhra Pradesh, India. *Journal of Vertebrate Paleontology* 14:260–77.

Prinn, R. G. and B. Fegley, Jr. 1987. Bolide impacts, acid rain, and biospheric traumas at the Cretaceous-Tertiary boundary. *Earth and Planetary Science Letters* 83:1–15.

Raup, D. M. 1991. *Extinction: Bad Genes or Bad Luck?*. New York: W. W. Norton.

Raup, D. M. and D. Jablonski. 1993. Geography of end-Cretaceous marine bivalve extinctions. *Science* 260:971–73.

Raup, D. M. and J. J. Sepkoski, Jr. 1982. Mass extinction in the marine fossil record. *Science* 215:1501–3.

——. 1984. Periodicity of extinctions in the geologic past. *Proceedings of the National Academy of Sciences, USA* 81:801–5.

Retallack, G. J. 1993. Evidence from paleosols for acid overdose at the end of the Cretaceous in Montana. *Journal of Vertebrate Paleontology Abstracts of Papers* 13, supplement to number 3, p. 54A.

——. 1994. A pedotype approach to latest Cretaceous and earliest Tertiary paleosols in eastern Montana. *Geological Society of America Bulletin* 106:1377–97.

Rigby, J. K., Jr., L. W. Snee, D. M. Unruh, S. S. Harlan, J. Guan, F. Li, J. J. Rigby, Sr., and B. J. Kowalis. 1993. 40Ar/39Ar and U-Pb dates for dinosaur extinction, Nanxiong Basin, Guongdong Province, People's Republic of China. *Geological Society of America Abstracts with Programs* 25 (6): 296.

Rounsevell, D. E. and S. J. Smith. 1982. Recent alleged sightings of the thylacine (Marsupialia, Thylacinidae) in Tasmania. In M. Archer, ed., *Carnivorous Marsupials*, vol. 1, pp. 233–36. Mosman, N.S.W., Australia: Royal Zoological Society of New South Wales.

Rowe, T. 1993. Phylogenetic systematics and the early history of mammals. In F. S. Szalay, M. J. Novacek, and M. C. McKenna, eds., *Mammal Phylogeny, Volume 1: Mesozoic Differentiation, Multituberculates, Montremes, Early Therians, and Marsupials*, pp. 129–45. New York: Springer-Verlag.

Rowe, T., R. L. Cifelli, T. M. Lehman, and A. Weil. 1992. The Campanian

Terlingua Local Fauna, with a summary of other vertebrates from the Aguja Formation, Trans-Pecos Texas. *Journal of Vertebrate Paleontology* 12:472–93.

Ruben, J. 1995. The evolution of endothermy in mammals and birds: From physiology to fossils. *Annual Review of Physiology* 57:69–95.

Russell, D. A. 1967. A census of dinosaur specimens collected in western Canada. *National Museum of Canada Natural History Paper* 36:1–13.

——. 1984. The gradual decline of the dinosaurs—fact or fallacy? *Nature* 307:360–61.

Savrda, C. E. 1993. Ichnosedimentologic evidence for a noncatastrophic origin of Cretaceous-Tertiary boundary sands in Alabama. *Geology* 21:1075–78.

Schopf, J. W. and M. R. Walter. 1983. Archean microfossils: New evidence of ancient microbes. In J. W. Schopf, ed., *Earth's Earliest Biosphere: Its Origin and Evolution*, pp. 214–39. Princeton: Princeton University Press.

Sepkoski, Jr., J. J. 1990. The taxonomic structure of periodic extinction. In V. L. Sharpton and P. Ward, eds., *Global Catastrophes in Earth History: An Interdisciplinary Conference on Impacts, Volcanism, and Mass Mortality*, Special Paper 247, pp. 33–44. Boulder, Colo.: Geological Society of America.

Sepkoski, Jr., J. J. and A. I. Miller. 1985. Evolutionary faunas and the distribution of Paleozoic benthic communities in space and time. In J. W. Valentine, ed., *Phanerozoic Diversity Patterns: Profiles in Macroevolution*, pp. 153–90. Princeton: Princeton University Press.

Sharpton, V. L., K. Burke, A. Camargo-Zanoguera, S. A. Hall, D. S. Lee, L. E. Marín, G. Suárez-Reynoso, J. M. Quezada-Muñeton, P. D. Spudis, and J. Urrutia-Fucugauchi. 1993. Chicxulub multiring impact basin: Size and other characteristics derived from gravity analysis. *Science* 261:1564–66.

Sheehan, P. M. and D. E. Fastovsky. 1992. Major extinctions of land-dwelling vertebrates at the Cretaceous-Tertiary boundary, eastern Montana. *Geology* 20:556–60.

Sheehan, P. M., D. E. Fastovsky, R. G. Hoffman, C. B. Berghaus, and D. L. Gabriel. 1991. Sudden extinction of the dinosaurs: Latest Cretaceous, upper Great Plains, USA. *Science* 254:835–39.

Signor, III, P. W. and J. H. Lipps. 1982. Sampling bias, gradual extinction patterns and catastrophes in the fossil record. In L. T. Silver and P. H. Schultz, eds., *Geological Implications of Impacts of Large Asteroids and Comets on the Earth*, Special Paper 190, pp. 291–96. Boulder, Colo.: Geological Society of America.

Sleigh, M. 1989. *Protozoa and Other Protists*. London: Edward Arnold.

Sloan, R. E., J. K. Rigby, Jr., L. M. Van Valen, and D. L. Gabriel. 1986. Gradual dinosaur extinction and simultaneous ungulate radiation in the Hell Creek Formation. *Science* 234:1173–75.

Sloan, R. E. and L. Van Valen. 1965. Late Cretaceous mammals from Montana. *Science* 148:220–27.

Smit, J. and G. Klaver. 1981. Sanidine spherules at the Cretaceous-Tertiary boundary indicate a large impact event. *Nature* 292:47–49.

Smit, J. and S. van der Kaars. 1984. Terminal Cretaceous extinction in the Hell Creek area, Montana: Compatible with catastrophic extinction. *Science* 223:1177–79.

Smith, A. B. 1994. *Systematics and the Fossil Record*. Oxford: Blackwell Scientific Publications.

Smith, A. G., A. M. Hurley, and J. C. Briden. 1981. *Phanerozoic Paleo-continental World Maps*. Cambridge: Cambridge University Press.

Smith, A. G., D. G. Smith, and B. M. Funnell. 1994. *Atlas of Mesozoic and Cenozoic Coastlines*. Cambridge: Cambridge University Press.

Smith, K. A. and N. T. Maffitt. 1994. Wildlife conservation law in California: From bag limits to biodiversity. In C. G. Thelander, ed., *Life on the Edge: A Guide to California's Endangered Natural Resources, Wildlife*, pp. 32–40. Santa Cruz, Calif.: BioSystems Books.

Smith, M. 1982. Review of the thylacine (Marsupialia, Thylacinidae). In M. Archer, ed., *Carnivorous Marsupials*, vol. 1, pp. 237–53.

Spencer, E. W. 1964. *Basic Concepts of Historical Geology*. New York: Thomas Y. Crowell.

Springer, V. G. and J. P. Gold. 1989. *Sharks in Question*. Washington, D.C.: Smithsonian Institution Press.

Stets, J., A.-R. Ashraf, H. K. Erben, G. Hahn, U. Hambach, K. Krumsiek, J. Thein, and P. Wurster. 1995. The Cretaceous-Tertiary boundary in the Nanxiong Basin (continental facies, southeast China). In N. MacLeod and G. Keller, eds., *The Cretaceous-Tertiary Mass Extinction: Biotic and Environmental Effects*. New York: W. W. Norton.

Sweet, A. R., D. Braman, and J. F. Lerbekmo. 1993. Northern mid-continental Maastrichtian and Paleocene extinction events. *Geological Association of Canada, Program and Abstracts*, p. 103.

Swisher, III, C. C., L. Dingus, and R. F. Butler. 1993. 40Ar/39Ar dating and magnetostratigraphic correlation of the terrestrial Cretaceous-Paleogene boundary and Puercan Mammal Age, Hell Creek–Tullock formations, eastern Montana. *Canadian Journal of Earth Sciences* 30:1981–96.

Swisher, III, C. C., J. M. Grajales-Nishimura, A. Montanari, S. V. Margolis, P. Claeys, W. Alvarez, P. Renne, E. Cedillo-Pardo, F. J.-M. R. Maurrasse, G. H. Curtis, J. Smit, and M. O. McWilliams. 1992. Coeval 40Ar/39Ar ages of 65.0 million years ago from Chicxulub crater melt rock and Cretaceous-Tertiary boundary tektites. *Science* 257:954–58.

Tipper, J. C. 1979. Rarefaction and rarefaction—the use and abuse of a method in paleontology. *Paleobiology* 5:423–34.

Toon, T. P., J. P. Pollack, T. P. Ackerman, R. P. Turco, C. P. McKay, and M. S. Liu. 1982. Evolution of an impact-generated dust cloud and its effects on the atmosphere. In L. T. Silver and P. H. Schultz, eds., *Geological Implications of Impacts of Large Asteroids and Comets on the Earth*, Special Paper 190, pp. 187–200. Boulder, Colo.: Geological Society of America.

Tschudy, R. H., C. L. Pillmore, C. J. Orth, J. S. Gilmore, and J. D. Knight. 1984. Disruption of the terrestrial plant ecosystem at the Cretaceous/Tertiary boundary, western interior. *Science* 225:1030–32.

Turco, R. P., O. B. Toon, T. P. Ackerman, J. P. Pollack, and C. Sagan. 1983. Nuclear winter: Global consequences of multiple nuclear explosions. *Science* 222:1283–92.

Vannote, R. L., G. W. Minshall, K. W. Cummins, J. R. Sedell, and C. E. Cushing.

1980. The river continuum concept. *Canadian Journal of Fisheries and Aquatic Science* 37:130–37.
Van Valen, L. 1978. The beginning of the Age of Mammals. *Evolutionary Theory* 4:45–80.
Van Valen, L. and R. E. Sloan. 1965. The earliest primates. *Science* 150:743–45.
——. 1977. Ecology and extinction of dinosaurs. *Evolutionary Theory* 2:37–64.
Vickers-Rich, P. and T. H. Rich. 1993. *Wildlife of Gondwana*. Chatswood, N.S.W., Australia: Reed.
Waage, K. M. 1968. The type Fox Hills Formation, Cretaceous (Maestrichtian), South Dakota. Part 1. Stratigraphy and paleoenvironments. *Peabody Museum of Natural History, Yale University Bulletin* 27:1–175.
Wang, K., M. Attrep, Jr., and C. J. Orth. 1993. Global iridium anomaly, mass extinction, and redox change at the Devonian-Carboniferous boundary. *Geology* 21:1071–74.
Ward, P. D. 1990. A review of Maastrichtian ammonite ranges. In V. L. Sharpton and P. Ward, eds., *Global Catastrophes in Earth History: An Interdisciplinary Conference on Impacts, Volcanism, and Mass Mortality*, Special Paper 247, pp. 519–30. Boulder, Colo.: Geological Society of America.
——. 1994. *The End of Evolution: On Mass Extinctions and the Preservation of Biodiversity*. New York: Bantam Books.
Weil, A. 1994a. Acid rain as an agent of extinction at the K/T boundary—not! *Journal of Vertebrate Paleontology Abstracts of Papers* 14, supplement to number 3, p. 51A.
——. 1994b. K/T survivorship as a test of acid rain hypotheses. *Geological Society of America Abstracts with Programs* 26 (7): 335.
Weishampel, D. B. 1983. Hadrosaurid jaw mechanics. *Acta Palaeontologica Polonica* 28:271–80.
——. 1990. Dinosaurian distribution. In D. B. Weishampel, P. Dodson, and H. Osmólska, eds., *The Dinosauria*, pp. 63–139. Berkeley: University of California Press.
Weishampel, D. B., P. Dodson, and H. Osmólska, eds. 1990. *The Dinosauria*. Berkeley: University of California Press.
Weishampel, D. B. and J. R. Horner. 1990. Hadrosauridae. In D. B. Weishampel, P. Dodson, and H. Osmólska, eds., *The Dinosauria*, pp. 534–61. Berkeley: University of California Press.
Wellenhofer, P. 1991. *The Illustrated Encyclopedia of Pterosaurs*. New York: Crescent Books.
Whetsone, K. N. 1978. A new genus of cryptodiran turtles (Testudinoidea, Chelydridae) from the upper Cretaceous Hell Creek Formation of Montana. *The University of Kansas Science Bulletin* 51:539–63.
Williams, M. E. 1994. Catastrophic versus noncatastrophic extinction of the dinosaurs: Testing, falsifiability, and the burden of proof. *Journal of Paleontology* 68:83–190.
Wilson, E. O. 1992. *The Diversity of Life*. Cambridge, Mass.: Harvard University Press.
Wilson, M. V. H., D. B. Brinkman, and A. G. Neuman. 1992. Cretaceous

Escoidei (Teleostei): Early radiation of the pikes in North American fresh waters. *Journal of Paleontology* 66:839–46.

Wilson, M. V. H. and R. R. G. Williams. 1994. Systematic position of the enigmatic teleost *Platacodon nanus* Marsh, from the Upper Cretaceous of North America. *Journal of Vertebrate Paleontology Abstracts of Papers* 14, supplement to number 3, pp. 52A–53A.

Wolbach, W. S., I. Gilmour, and E. Anders. 1990. Major wildfires at the Cretaceous/Tertiary boundary. In V. L. Sharpton and P. Ward, eds., *Global Catastrophes in Earth History: An Interdisciplinary Conference on Impacts, Volcanism, and Mass Mortality*, Special Paper 247, pp. 391–400. Boulder, Colo.: Geological Society of America.

Wolfe, J. A. and G. R. Upchurch, Jr. 1986. Vegetation, climatic, and floral changes at the Cretaceous-Tertiary boundary. *Nature* 324:148–52.

Zinsmeister, W. J. and R. M. Feldmann. 1994. Antarctica, the forgotten stepchild: A view of KT extinction from the high latitude southern latitudes. New developments regarding the KT event and other catastrophes in earth history. *Lunar and Planetary Institute Contribution*, no. 825: 134–35.

Zinsmeister, W. J., R. M. Feldmann, M. O. Woodburne, and D. H. Elliot. 1989. Latest Cretaceous/earliest Tertiary transition on Seymour Island, Antarctica. *Journal of Paleontology* 63:731–38.

Index

References to figures appear in *italic* type; references to tables appear in **bold** type.